albatross

their world, their ways
albatross

Tui De Roy   Mark Jones   Julian Fitter

FIREFLY BOOKS

A FIREFLY BOOK

Published by Firefly Books Ltd. 2008

Copyright © 2008 Tui De Roy, Mark Jones and Julian Fitter
Typographical design copyright © 2008 David Bateman Ltd.
Additional contributors retain copyright to their individual texts and illustrations.

All photographs are by Tui De Roy (unless credited otherwise).
They have been taken in wild and free conditions and images have not been digitally altered or otherwise manipulated.

All rights reserved. No part of this publication may be reproduced, stored in a retrieval system or transmitted in any form or by any means, electronic, mechanical, photocopying, recording or otherwise, without the prior written permission of the publisher.

Publisher Cataloging-in-Publication Data (U.S.)
De Roy, Tui.
    Albatross : their world, their ways / Tui De Roy ; Mark Jones ; Julian Fitter.
[240] p. : col. photos., maps ; cm.
Includes index.
Summary: Comprehensive book on all albatross species, featuring the latest writings and
    research from international experts accompanied by color photographs.
ISBN-13: 978-1-55407-415-0
ISBN-10: 1-55407-415-0
1. Albatrosses.   I. Jones, Mark.   II. Fitter, Julian.   III. Title.
598.42 dc22   QL696.P63.D476 2008

Library and Archives Canada Cataloguing in Publication
De Roy, Tui
    Albatross : their world, their ways / Tui De Roy, Mark Jones, Julian Fitter.
ISBN-13: 978-1-55407-415-0
ISBN-10: 1-55407-415-0
1. Albatrosses — Pictorial works. 2. Albatrosses — Identification.
I. Jones, Mark   II. Fitter, Julian   III. Title.
QL696.P63D47 2008   598.42   C2008-900756-5

Published in the United States by
Firefly Books (U.S.) Inc.
P.O. Box 1338, Ellicott Station
Buffalo, New York 14205

Published in Canada by
Firefly Books Ltd.
66 Leek Crescent
Richmond Hill, Ontario L4B 1H1

Design: Trevor Newman
Maps: Tim Nolan/Black Ant

Printed in China

Page 1 Northern royal albatross; *pages 2–3* Atlantic yellow-nosed albatross, Tristan da Cunha Island, South Atlantic; *pages 4–5* Antipodean albatross, Adams Island, New Zealand; *pages 6–7* Light-mantled albatross, Campbell Island, New Zealand; *pages 8–9* Waved albatross, Española Island, Galapagos, Ecuador; *this page* Antipodean albatross, Adams Island, New Zealand; *contents page* Southern royal albatross, Campbell Island, New Zealand.

Contents

	Foreword HRH Prince Charles	14
	Map of Global Distribution of Albatross	16
	Saving Albatrosses Around the World	18
	Acknowledgements	19
	Introduction Carl Safina	20

Part One: Spirits of the Oceans Wild *Tui De Roy*

1	Storm Riders of the Southern Ocean: The Wanderers	24
2	Racing Royalty: The Royals	40
3	Travellers of Wind and Wave: The Black-Brows	52
4	Dynamic Wings and Painted Bills: Grey-headed, Yellow-nosed and Buller's	62
5	Down Under Specials: The Shy Tribe	78
6	Enigmatic Elegance: The Sooties	96
7	North Pacific Survivors: The Northern Albatrosses	110
8	Under the Tropical Sun: The Galapagos Albatross	124

Part Two: Science and Conservation *Mark Jones*

9	Perspectives: Albatrosses and Man through the Ages *Mark Jones*	138
10	Flagship Species at Half-mast *Rosemary Gales*	148
11	Albatross Flight Performance and Energetics *Scott Shaffer*	152
12	Do Wanderers Always Return? *Michael Double*	154
13	Albatross Populations and Migrations: From Observation to Application *John P. Croxall*	156
14	Indian Ocean Albatrosses: A Status Report *Henri Weimerskirch*	158
15	Waved Albatross *David Anderson*	160
16	Short-tailed Albatross: Dandies of the Deep *Greg Balogh*	162
17	Circumpolar Royal Travellers *Christopher J.R. Robertson*	164
18	Buller's Albatross *Paul Sagar*	166
19	Chatham Albatross *Paul Scofield*	168

20	The Plight of the Albatross on Tristan da Cunha *Conrad J. Glass*	170
21	Southern Seabird Solutions Trust: Conservation through Cooperation *Janice Molloy, John Bennett, Caren Schroder*	172
22	The Albatross Task Force: A Sea Change for Seabirds *Ben Sullivan*	174
23	South American Perspective: Fisheries Mortality *Marco Favero*	176
24	Conserving Magnificent Flyers: A Personal Journey with Southern Albatrosses *John Cooper*	178
25	Applying Spatially-explicit Measures for Albatross Conservation *David Hyrenbach*	180
26	Conversation with an Ex-High Seas Poacher *Tui De Roy*	182

Part Three: Species Profiles *Julian Fitter*

Introduction to Albatrosses, Mollymawks and Gooneys	186
Species Profiles	192
Where to See Albatrosses	233
Glossary: Selected Terms and Abbreviations	234
Further Reading	237
Index	238

Foreword
HRH Prince Charles

ABOVE His Royal Highness Prince Charles, here seen visiting the northern royal albatrosses of Tairoa Head in New Zealand, is an active supporter of BirdLife International's Save the Albatross campaign, often speaking passionately of our need to ensure their survival.
PHOTOGRAPH COURTESY OTAGO DAILY TIMES/PHOTOGRAPHER CRAIG BAXTER

RIGHT AND OPPOSITE Antipodean albatrosses frequent the rich waters of the 1000-metre (3280 ft) deep undersea Kaikoura Canyon along the east coast of New Zealand, a favourite feeding ground.

CLARENCE HOUSE

Albatrosses are iconic creatures, a flagship species for the conservation of the oceans as a whole, just as rhinos, pandas and tigers have become for the land. They are extraordinary, almost mythical creatures, with their enormous wingspan, great longevity and remarkable powers of ocean navigation and travel, almost transcending the very concept of what it means to be a bird. I remember so well when I was in the Royal Navy standing on the deck of a fast-moving warship in one of the Southern oceans, watching an albatross maintaining perfect position alongside for hour after hour, and apparently day after day. It is a sight I will never forget, and I find it unthinkable that we could extinguish them for ever, never to be resurrected. But unless action is taken, that is _exactly_ what will happen.

The plight of the albatross should remind us of the ultimate fragility of all the migratory species that mark the great cycle of the seasons and the mysterious, inner unseen urge that compels such creatures to follow, with unerring accuracy, the timeless patterns of movement around this globe. But they are now dependent upon the whim of man – either we can choose to do something to save them, or stand by and let them disappear. If we allow this to happen then I, for one, think we would sacrifice any claim whatsoever to call ourselves civilized beings. We will have violated something profoundly sacred in the inner workings of Nature, and future generations will never forgive us.

Charles

Global Distribution of Albatross

Briesemeister projection (oblique azimuthal equal-area) centred on 50°S 160°E

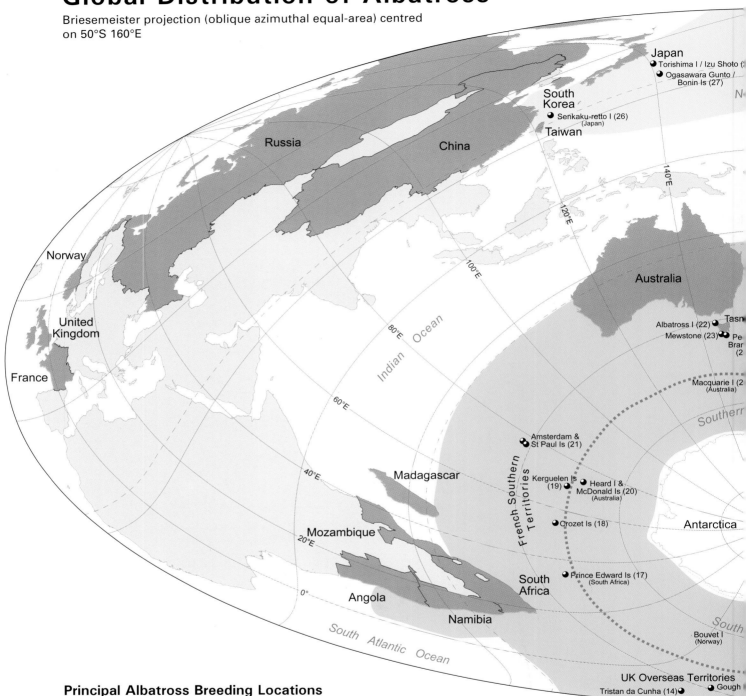

Principal Albatross Breeding Locations
Island numbers referenced on map (italics denotes main colonies):

Wandering   *14, 17, 18, 19,* 25
Antipodean   *1, 2,* 5
Tristan   15, *16*
Amsterdam   *21*
Northern royal   *6,* 8
Southern royal   *1,* 2
Black-browed   1, 5, *10, 11, 12, 13, 14,* 18, *19,* 20, 25
Campbell   *1*

White-capped   *2,* 5
Shy   *22, 23,* 24
Salvin's   3, *4*
Chatham   *6*
Grey-headed   *1, 12,* 14, *17, 18, 19,* 25
Buller's   3, *6, 7,* 9
Atlantic yellow-nosed   15, *16*
Indian yellow-nosed   *17, 18,* 19, *21*

Light-mantled   1, 2, 5, 14, 17, 18, 19, 20, 25
Sooty   15, 16, 17, 18, 21
Waved   32, 33
Laysan   27, 29, 30, 31
Black-footed   26, 27, 28, 29
Short-tailed   26, 28

Saving Albatrosses Around the World

The greatest avian nomads of the oceans are in trouble. They are the victims of many human-induced factors, such as ocean pollution, climate change and alien predators introduced to their breeding islands. But, most of all, they are dying due to the activities of high seas fishing fleets. As a result, almost 90 per cent of all albatross species are either declining or otherwise at risk of extinction, making them the most threatened of any multi-species bird family.

Awareness is growing, and more and more people and organisations around the world — governmental and non-governmental, business and private — are working to reverse this alarming trend. Much remains to be done, but the momentum is building.

The organisations whose logos and websites appear below all share in the vision of saving the world's albatrosses. *Note*: The use of these logos does not imply that the organisations represented endorse this book or its contents.

www.blueocean.org

www.avesconservacion.org

www.birdsaustralia.com.au

www.iucn.org

www.falklandsconservation.com

www.southernseabirds.org

www.noaa.gov

www.fws.gov

www.acap.aq

www.doc.govt.nz

www.projetoalbatroz.org.br

www.rspb.org

www.anu.edu.au

www.prodelphinus.org

www.abcbirds.org

www.forestandbird.org.nz

www.aad.gov.au

www.ccamlr.org

www.darwinfoundation.org

www.pacificseabirdgroup.org

www.mundoazul.org

www.wsg.washington.edu

www.niwa.co.nz

www.galapagospark.org

Albatrosses are majestic and iconic symbols of freedom and the open oceans. However, many populations are declining at an alarming rate. Through the Save the Albatross campaign, we have been working since 1997 to raise awareness of the plight of albatrosses and advocate at international, national and local levels for the adoption of appropriate regulations and practical mitigation measures to reduce albatross mortality. BirdLife International acknowledges this book as an important contribution towards these goals.

Michael Rands
Director and Chief Officer
www.birdlife.org

This book is endorsed by the International League of Conservation Photographers (ILCP), an initiative of the WILD Foundation, as a valuable conservation tool developed in synergy with BirdLife International's Save the Albatross Campaign. It aligns with the ILCP's core values of using credible, compelling visually based messages, while adhering to the highest ethical standards, to address one of the critical issues of concern in the biodiversity of our oceans.

Cristina Mittermeier
Executive Director
www.ILCP.com

Acknowledgements

Like the albatrosses themselves, who range unseen over vast regions of our globe, so the people who have in myriad ways contributed to the creation of this book hail from many diverse and remote parts. And because this project has evolved over many years, the list of people who have come to influence its successful outcome would be too long and complex to enumerate in full. As a result, we wish to thank not only those mentioned in this limited space, but all those who share in the vision of a world safe for albatrosses to continue flying free forever.

At the concept level, we are grateful to Paul Sagar of NIWA, Michael Rands of BirdLife International, Pete McClelland and Greg Lind of the New Zealand Department of Conservation and the many others who gave the idea life, and to Paul Bateman and Tracey Borgfeldt, our publishers, for believing in its potential from start to finish. We thank Carl Safina for giving us inspiration and writing a superb introduction. Special thanks to Julie Cornthwaite for her considerable input, from organising and proofreading to liaising with contributors. Likewise Roz Kidman Cox for her invaluable insight and advice with the photo selection.

We have been overwhelmed by the enthusiastic response from researchers and other experts around the world who shared their knowledge through essay contributions and illustrations. We wish to thank them for giving so freely of their time: Rosemary Gales, Scott Shaffer, Mike Double, John Croxall, Henri Weimerskirch, Dave Anderson, Greg Balogh, Chris Robertson, Paul Sagar, Paul Scofield, Connie Glass, Janice Molloy, John Bennett, Caren Schroder, Ben Sullivan, Marco Favero, John Cooper, David Hyrenbach, Gary Drew and D.C. Nichols. We are furthermore indebted to those who helped us out with photos of the three albatross species we did not manage to cover ourselves, plus some other exceptional images: Peter Ryan, Barry Baker, Dennis Coutts, Ed Melvin, Richard Cuthbert, Jeffrey Mangel, Crissy Wickes, Rosemary Gales, Scott Shaffer, Matt Charteris, Marc Romano, Mark Royo Celano, Sarah Crofts, Andrea Angel and Ross Wanless.

Dave Anderson, Jill Awkerman, Rob Suryan and, especially, Paul Sagar meticulously reviewed the species facts in Part Three.

A number of people have provided steadfast support in more ways than we can possibly describe, from our first albatross encounters throughout the complex logistics that ultimately let our project take flight, among them Colin Monteath, Meri and Ian Leask, Henk Haazen, Peter Gaze, Jeremy Carroll, Peter Moore, Doug Veint, Charles Hufflett, Rodney and Shirley Russ, Rowley Taylor (New Zealand); Mike Harris, Harry Scott, Alex Jones (UK); Henry Valentine, Dorrien Ven, Capt. Tad de Oliveira, Capt. Clarence October, Capt. Peter Warren (South Africa); Mike and Janice Hentley, Jimmy and Felicity Glass, Eric McKenzie, Claire Volkwyn (Tristan da Cunha); Kim Rivera, Beth Flint, Ron leValley, Mark Rouzon, Cristina Mittermeier (US); Dolores Diez, Godfrey Merlen, Eliecer Cruz, Felipe Cruz, Rob Bensted-Smith (Galapagos); Phil and Stella Middleton, Roddy Napier, the Pole-Evans family, Tony Chater, Ian and Maria Strange, Sally Poncet (Falkland Islands); Capt. Victor Vassiliev, Capt. Peter Golokov (Antarctica); Quark Expeditions, Lindblad Expeditions. Special mention goes to Ada Hough and Hariroa Daymond of the Chatham Islands iwi (indigenous tribes), who demonstrated exceptional trust in granting us access to places almost never visited.

Finally, there are the very special people who partake in our albatross passion and with whom we've shared memorable slices of life, whether sailing aboard our small yacht *Mahalia* through southern storms or spending time together on wild and wondrous albatross islands: Peter Fullerton, Jacinda Amey, Grant Redvers, Carl Harmer, Chris Rickard, Sarah Frazer, Andy Clark, Alison Ballance, David Anderson, Kate Huyvaert, Brian Bowie and his Gough Island team, Paul Scofield, Kennedy Warne, Paul Sagar, Jean Claude Stahl, Carl Safina, Colin O'Donnell, Jane Sedgely. Thank you all for sharing the journey. And, finally, thank you Alan, Julie and Jayne for flying with us on albatross wings.

ABOVE A pair of waved albatrosses on Española Island in the Galapagos bond tenderly at the onset of the nesting season.

Introduction
Carl Safina, adapted from *Eye of the Albatross*

The spirit who bideth by himself
In the land of mist and snow
He loved the bird that loved the man
Who shot him with his bow.

Dr. Carl Safina brought ocean conservation into the environmental mainstream. His 100+ publications and award-winning books include Song for the Blue Ocean, Eye of the Albatross, and Voyage of the Turtle. Recognition for his work includes a Pew Fellowship, Lannan Literary Award, John Burroughs Medal, and a MacArthur Prize. www.blueocean.org

BELOW Riding the raging winds of the latitudes known as the Roaring Forties, a southern royal albatross returns from a foraging trip to its breeding grounds on Campbell Island in the New Zealand Subantarctic.

In Samuel Taylor Coleridge's 1798 epic poem, *The Rime of the Ancient Mariner* (part of which is quoted above), the sailor who kills an albatross is compelled to wear around his neck the evidence of his crime against nature. Even two centuries ago, the bird symbolised beneficent companionship, harmed only at our peril. Coleridge, who never saw an albatross, sensed that here was a seabird with power enough to convey a universal cautionary tale. We sense it still.

Coleridge was not the only person who felt the symbolic power of albatrosses. Even scientists repeatedly freighted the birds with metaphor and meaning, draping them with everything from heroic virtue to fear and foreboding. Thus the genus name scientifically denoting most albatrosses is the Latin *Diomedea*. Diomedes was one of Homer's war heroes. During one campaign he so offended the goddess of wisdom, Athene, that in retribution she beset his fleet with a terrifying storm. When, rather than acting contrite, some of his crew further taunted the goddess, she transformed them into large white birds, 'gentle and virtuous'. The genus of dark-plumaged sooty albatrosses, *Phoebetria*, derives from the Latin, *phoebetron*, an object of terror, and the Greek *phoibetria*, a prophetess or soothsayer. *Exulans*, the wandering albatross's specific name, means 'out of one's country', thus to live in exile. That may be how the mariners on multi-year voyages who saw them felt. But the albatrosses themselves were always quite at home.

In the metaphors we make of the creatures of the heavens and the deep, we often project our imagery, imbuing them with our own reflection. But the world is more than a colouring book of shapes for us to fill in. When we perceive metaphor in reality we enhance our understanding of ourselves. But when we install meanings instead of seeing reality, we miss all the true texture and inherent value, like a child doodling over a great masterpiece. Loading up an albatross with our own symbols is a bit unfair — unfair to the animal, who suffers the bias of impressions we've created, and unfair to us; we miss the expansive opportunity of knowing other creatures. Why force albatrosses to wear humans around their necks? Sometimes an albatross is simply a bird. When we see that, worlds open.

These immense creatures we call 'albatross' are the greatest long-distance wanderers on earth. Big birds in big oceans, albatrosses lead big, sprawling lives across space and time, travelling to the limits of seemingly limitless seas. They accomplish these distances by wielding the impressive — wondrous, really — body architecture of creatures built to glide indefinitely.

An albatross is a great symphony of flesh, perception, bone and feathers, composed of long movements and set to ever-changing rhythms of light, wind, water. The musicality of an albatross in air derives not just from the bird itself but from the contrapuntal suite of action and inaction from which it composes flight. The creature drifts in the atmosphere at high speed, but itself remains immobile — an immense bird holding stock-still yet shooting through the wind. Just as individual notes become music by relationship to other notes, the bird's stillness becomes movement by context. Following your travelling ship with ease, watching you, circling stern to prow and back at will, it flies with scarcely a flinch, skimming wave upon wave, mile after mile after mile. Watching it, you invariably wonder, 'How can it do that?'

Exerting no propelling power of its own over long distances, it is driven by the tension between the

LEFT Antipodean albatross, Kaikoura, New Zealand.

two greatest forces on our planet: gravity and the solar-powered wind. The huge bird's placid mastery of gales never fails to impress mariners distressed by heavy weather. Charles Darwin, in a tempest near Cape Horn while aboard the *Beagle* in 1833, wrote, 'Whilst we were heavily laboring, it was curious to see how the Albatross … glided right up the wind.' Not far from there a few years ago, in a storm so great it stopped our 270-foot ship for a day, I too watched wandering albatrosses somehow gliding directly into 70-knot winds in hurricane conditions, circling our paralysed ship with surreal serenity, seeming oblivious to the shrieking, spume-filled gusts.

While mariners marvelled at the sheer size and stamina of albatrosses for centuries, the birds' oceanic travels were impossible to cipher. Where did they go? Sailors speculated, and some came close. Scientists guessed wrong. No one could have fully imagined, because albatrosses exert almost unimaginable lives. In the last few years albatrosses have been tracked by earth-orbiting satellites, and their true travels outdistance all previous conjecture.

During their whole lifespan they expend 95 per cent of their existence at sea — flying most of that time. Before maturing, albatrosses remain at sea for years, never alighting upon a solid surface, perhaps not glimpsing land, all the while. Theirs is a fluid world of wind and wild waters, everything in perpetual motion. When they do breed, albatrosses haunt only the most removed islands, hundreds, sometimes thousands of miles from any continent. And even at the most isolated island groups, albatrosses often choose to nest on the tiniest offshore islets, as though they can barely tolerate land at all.

But even living so far from humanity, albatrosses increasingly share a human-dominated destiny. Because they range so far and live so long, albatrosses contend with almost every effect that people exert upon the sea. From the elemental world of wind and water, the albatross's realm has come to encompass every complexity from fishing boats to chemical alterations to human-caused climate changes. Everything people are doing to oceans, albatrosses feel.

As we've entered a new human-dominated era in our planet's history — from the Holocene into the Anthropocene — albatrosses take on a new metaphorical role: they are one measure of our success; their health is our health, their failure is our failure. No study of nature can now avoid confronting the great changes humans have wrought. Albatrosses need us to protect them from ourselves. We need albatrosses because many of us harbour an albatross within. Beneath the daily routine, our truer nature is this wandering spirit on expansive wings, hungering for a chance to search new horizons, to hurtle along with the wind, taking chances, taking the world as is comes, making tracks that will endure only in our memory, forming our personal map of life and time.

The albatross cannot seem to escape its own symbolic power. And any creature so powerfully superlative is a creature we cannot abide to lose. The challenges to albatross survival are great, perhaps overwhelming for some species. But solutions are coming, and that is cause for hope. Let's define hope as the belief that things can get better. But to believe that things can get better, then not act on that belief, reduces us to mere wishing. What happens to albatrosses, as with everything, depends on what all of us, as individuals, decide to do. One must have hope — and one must act.

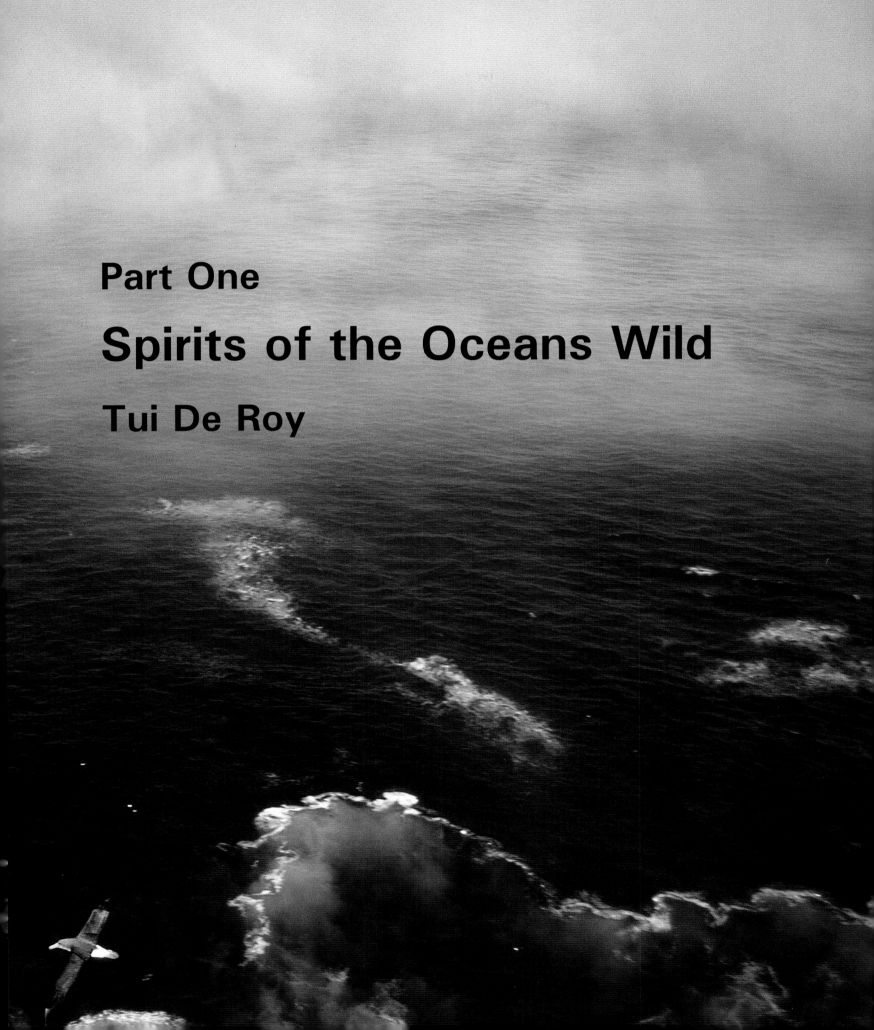

Part One
Spirits of the Oceans Wild
Tui De Roy

1
Storm Riders of the Southern Ocean
The Wanderers

Wandering albatross

Through the roaring wind and lashing salt spray I strain to keep my eyes on a white speck rising and falling in rhythmic arcs over the blurred horizon. Tracing a see-sawing path, yet travelling towards a fixed goal through the punishing South Atlantic gale, the speck soon draws near. A splendid male wandering albatross, recognisable by the heavy set of his pink bill, draws abreast of the pitching icebreaker labouring through the giant swells. Riding on immaculate, rigid wings spanning well over three metres (10 ft), his mastery of the elements is absolute.

All around, the maelstrom rages. Huge swells curl and break. Waves crash and collide, spume torn from their crests. The air is thick with droplets of stinging sea water flung high by shrieking gusts. In contrast with this bewildering tumult, the great bird is completely serene as he passes. Where everything around him is chaos, he seems to hang almost motionless for long instants, wings taut, feathers smooth and streamlined, his entire body a perfection of aerodynamics.

I seek relief from the elements by cowering behind the bridge structure of the ship, while the albatross cruises steadily by. His dark eye scrutinises me briefly, then his sailplane wings tip imperceptibly and he plunges towards the next wave trough, gaining speed. He is practising that marvel of all marvels in the realm of flight: harnessing the combined forces of gravity and wind to ride along with minimum energy expenditure of his own, a process described as dynamic soaring (see also Chapter 11). The trick, so fascinating to watch, is beautifully simple.

Dynamic soaring

Strong solar-powered winds, circulating freely around the Southern Ocean, where they meet no land in their easterly rush, whip up jagged waves

ABOVE Sometimes referred to as the snowy albatross, a pair of wandering albatrosses on South Georgia's Albatross Island, a major stronghold for the species, relax together at the nest before exchanging incubation duties.
OPPOSITE Gliding high over the storm-carved Windward Islets, Antipodean albatrosses return from distant wanderings to the island after which they are named.
PREVIOUS PAGES Critically endangered Chatham albatrosses soar through the mist around their sole nesting island, The Pyramid.

Spirits of the Oceans Wild • 25

ABOVE Nesting wandering albatrosses loosely sprinkled on the boggy plateau of Prince Edward Island form the densest colony of this species in the world; Atlantic yellow-nosed albatrosses nest on the cliffs beyond. Photo courtesy of Peter Ryan.

RIGHT A wandering albatross soars over the windy Southern Ocean. Photo: Mark Jones.

that roll along like liquid mobile mountains. This creates sufficient turbulence along the sea-air interface that the drag effectively slows the lower air layers even while the wind continues to travel at full speed higher up. The art in the albatross's flight is to extract lift from the airflow differentials by sailing up and down through these wind-speed gradients.

At first he glides smoothly in the trough of the wave, closely hugging the face of the oncoming swell until he begins to lose momentum. Then suddenly, with only the smallest of steering adjustments, he veers upwards over the top of the crest, rising sharply into the teeth of the wind. With taut wings reaching a nearly vertical plane, he gains a sudden velocity boost as he climbs steeply into the fastest air lanes 10 or 15 metres (35–50 ft) above the sea. Here he levels off, briefly embracing the air on arched wings until plunging once more downwind towards the sea.

For a few moments he follows the contour of the

ABOVE A male wandering albatross carries clumps of grass to add to the nest before taking over incubation from its mate. Albatross Island, South Georgia.

wave so closely his quivering wingtip may trace a thin, evanescent line upon the surface before taking another ride down the trough … then up again … repeating the entire performance in one contiguous, fluid sweep. Never does a feather move or flutter out of place and the wing adjustments are so precise, they are imperceptible to the human eye. Only his head tilts apace with his yo-yo flight so that his gaze remains steady with the horizon. Incredibly, researchers have found that albatrosses engaged in dynamic soaring expend no more energy than paddling on the sea surface.

Aside from his classic dark mottled wings, he is one of the whitest wandering albatrosses I've seen, with only fine filigree markings on his breast and shoulders. No doubt he belongs to the southernmost and largest of the four closely related species of Wanderers*, sometimes referred to as the snowy albatross.

The epitome of freedom and exuding utter control in one of the harshest environments on earth, this magnificent ocean traveller easily overtakes the ship and continues on his way. He rises and falls, rises and falls, with such cadence that he creates through my sense of vision an impression similar to that which a gentle waltz would impart through sound. Every time his wings veer upright, I remind myself incredulously that a tall basketball player with arms outstretched would barely reach the same span.

This regal bird seems to be on a mission, cruising an exceptionally steady course. For what seems like a long while I continue to watch him in the distance, but in fact he takes only a few minutes to vanish over the opposite horizon from whence he came, towards the large subantarctic island of South Georgia, still many hundreds of kilometres ahead.

South Georgia

Two days after this encounter an island appears between dark squalls, magnificent ramparts of ice

* 'Wandering albatross' can refer to the individual species *Diomedea exulans* and the group of four similar species, wandering, Amsterdam, Tristan and Antipodean, that were originally considered as one. To avoid confusion we have distinguished between the two by using a capital 'W' when referring to the group, as in 'the Wanderers' and a lower case 'w' when referring to the now recognised individual species, wandering albatross, *Diomedea exulans*.

Spirits of the Oceans Wild • 27

ABOVE Their plumage much darker than their more southerly relatives, two female Antipodean albatrosses show great interest in a displaying male on the misty plains of Antipodes Island.

and rock rising luminescent through curtains of sleet. More albatrosses soar across silvery pools of sunlight as our ship passes massive icebergs in the lee of the land approaching our anchorage. This is albatross country at its best. Within an hour I am wending my way past angry breeding bull fur seals towards the grassy top of Albatross Island.

And there he is. Highly unlikely of course, but at least in my mind's eye this could well be the very same albatross that passed us in the storm, come home to his mate sitting on the nest. While we cannot tell one albatross from another, they recognise each other implicitly. He stands here bolt upright, still imposing in size but somewhat stripped of his grace without the perpetual motion that characterised him at sea. Teetering slightly as though trying to adjust to being stationary, he takes a few steps and plucks some tufts of grass to add to the nest, his huge webbed feet — larger than my hands — slapping the mossy ground as he walks. He approaches the nest nodding and crooning gently, and his mate responds in kind. Together they tuck the new grass into their large nest mound, then delicately preen each other's fine facial feathers with massive hooked beaks.

These same two birds have probably met unfailingly to nest here every other year, during a lifetime that might last 60 or more years. Everything about the wandering albatross is on a scale that is hard for us to comprehend. Their huge single egg takes two and a half months to hatch, and their chick, which they will take turns feeding through the bitter winter, will not fledge until about this time next year. When the chick is small, feeding trips may be short at first — just a couple of days and a mere few hundreds of kilometres distant — but by the time it is a few months old mother and father may each travel for a month at a time, covering as much as 15,000 kilometres (over 9000 miles) on a single trip to bring food home. If they are successful in

raising this single youngster without mishap until it departs the island on its own wings, they must then take a year off before returning to nest again. And during that year it's anybody's guess how far they may wander.

Albatross travels

In 1887 a dead wandering albatross washed up on an Australian beach with a message inscribed on a rough metal band around its neck, telling of 13 shipwrecked sailors stranded on the Crozet Islands, nearly 5000 kilometres (3000 miles) to the west. The bird had left the islands just 45 days earlier. Taking far longer than that to send a search party, the would-be rescuers found only an abandoned camp, but this was the first indication of the extraordinary travels that wandering albatrosses routinely undertake. Satellite telemetry has since provided us with far greater insights, but the results remain astounding. Some albatrosses appear to triangulate

ABOVE Facing the prevailing westerly weather, nesting Antipodean albatrosses are thinly dispersed amongst the wind-scoured tussock grass overlooking the complex volcanic landscape of Antipodes Island, where nearly the entire world population breeds.
LEFT A pair of wandering albatrosses at South Georgia celebrate their reunion by adding more grass to their nest.

ABOVE While opportunistic pintado petrels dash in for scraps, two Antipodean albatrosses tussle violently over fish offal near the Kaikoura Peninsula of New Zealand, their demeanour at sea far more assertive than ashore at the nest. Photo: Mark Jones.

the open ocean with such precision they return to known haunts time and again, whereas others may wander freely from one ocean basin to another and back again, undeterred by the prevailing westerly winds. Here they may encounter close cousins who were once thought to belong to the same species, but with whom they never interbreed because each returns to its natal island to seek a mate, often for life. Because of this genetic isolation DNA studies now tell us that they are in fact distinct species (see Chapter 12).

Antipodean albatross

High on the fog-shrouded plateau of Antipodes Island, south-east of New Zealand, I meet the smallest of the Wanderer cousins, the Antipodean albatross. Mottled brown — or in the case of females, almost as dark as first-year juveniles — their smaller size seems to make them more nimble and active than their snowy cousins. My first visit is in spring when large chicks are just losing their winter down and furiously exercising newfound wings, aware that life will very soon depend totally on learning to fly. Tired parents visit only briefly with final deliveries of food to send them on their way. Chattering parakeets incongruously fossick around them among pink-flowered *Anesotome* megaherbs, plants belonging to the celery family.

Courtship season on Antipodes Island

Three months later I return to find the island transformed. The flower stalks are dry and so are the carcasses of chicks who had not built up enough energy reserves to leave. Yet, under low glowering skies, every grassy valley and dale resonates with the ecstatic cries of courting groups of young adults. Gathering in flocks of up to a dozen — called gamming groups after the habits of early sailors to get together and swap yarns — chocolate-brown females and somewhat paler males mingle

ABOVE An efficient scavenger used to scouring vast tracts of ocean in search of food, an Antipodean albatross makes a quick meal of a dead cookie cutter shark floating at the surface. Kaikoura Peninsula, New Zealand. Photo: Mark Jones.

LEFT A nine-month-old Antipodean albatross moults from down to feathers, preparing to leave the island with the arrival of summer.

Spirits of the Oceans Wild • 31

ABOVE A Gibson's (Antipodean) albatross, gliding along the towering volcanic cliffs of Adams Island, uses a southerly breeze to rise several hundreds of metres above the sea to the grassy slope where the nesting colony is located.

enthusiastically in a process that may take years before a firm pair bond is established. Returning to their natal island after six or seven years of ocean wanderings, these youngsters' excitement contrasts sharply with the total calm of this year's new breeding shift of stoically incubating mature birds.

Gibson's albatross

Several hundred miles to the south, on Adams Island in the Auckland Islands group, a similar scene takes place among the Gibson's albatross. With paler mottled plumage than those on Antipodes, the jury is still out as to whether the Gibson's is to be regarded as a species of its own, or a lighter coloured brother of the Antipodean. High above the breathtaking cliffs overlooking the immensity of the Southern Ocean, the subadult birds perform their late-summer pageantry trying to decide who to settle down with for their first

nesting attempt around the age of 12 or 13. With all the gusto of youth, they dance and prance, bow, bill clap and sky-point. The apex of their exuberant choreography is reached when they suddenly spread their magnificent wings wide as if to embrace their partners, sometimes twirling around one another in a stupendous pas de deux — or trois. Their island is one of the most pristine in the whole of the subantarctic region, having never suffered the invasion of any of the bevy of introduced species so often released from early sailing ships. Stringently protected by New Zealand's Department of Conservation, only occasional science parties are permitted to visit.

Tristan albatross

As far as you can get from New Zealand — in fact, on the opposite longitude — lie the foreboding, misty volcanoes of Tristan da Cunha

and Gough Island in the South Atlantic, home to another Wanderer relative, the Tristan albatross. Intermediate in size and shadings between the giant snowy Wanderer of the far south and the darker and smaller Antipodean, this species also shows the greatest colour difference between sexes of any albatross, the male being almost totally white whereas the female remains mottled brown throughout life.

Tristan da Cunha Island

Sadly, the Tristan albatross has vanished from the island after which it was named, having once served as a staple food supply for the small group of hardy pioneers who settled here nearly two centuries ago. Native-born James Glass, direct descendant of the Scottish founding father of the settlement, helps me explore the rugged interior of this massive but young volcano. We first follow a narrow zigzag path up a near-vertical 750-metre

ABOVE A pair of Gibson's (Antipodean) albatrosses greet each other as the sun sets over Adams Island, a pristine island in the Auckland group which has no record of introduced foreign organisms. Photo: Mark Jones
LEFT As a new nesting season begins on Adams Island and breeders are sitting on eggs, a trio of younger birds perform their ecstatic display.

Spirits of the Oceans Wild • 33

ABOVE LEFT A Tristan albatross chick sits in the spring sunshine on Gough Island moulting the last of its winter coat.
ABOVE RIGHT A returning female Tristan albatross, typically much darker than her mate, walks towards her nest across the waterlogged, mossy plateau of Goneydale on Gough Island.
FAR RIGHT Having survived winter on Gough Island, a young Tristan albatross prepares for its first flight.

(2460-ft) wall, with plunging views of the picturesque little settlement along the shoreline evocatively named Edinburgh of the Seven Seas. It is sobering to think that such a small human presence could have such dire effects on an entire albatross species. But James, who heads the island government's Natural Resources Department and looks after conservation matters, explains how the islanders' attitudes have completely changed and all nesting seabirds are now protected by law (see Chapter 20). To his deep satisfaction several of the great birds have recently been seen prospecting on the volcano's sloping plateau, and there is hope that they will eventually recolonise these ancestral grounds.

Following mist-clad knife-edge ridges and crossing massive fog-filled gulches, we erupt into brilliant sunshine approaching the steep 2060-metre (6758-ft) summit. Luscious moss cushions give way to glassy black scoria fields streaked with summer snow banks. Gazing over an ocean of clouds it seems I should be able to see Africa to one side and South America to the other, so vast is the feeling of endless space. Far, far below I know the great white birds are skimming the waves on their way to another island refuge.

Gough Island

An overnight sail on a cray-fishing vessel takes me to Gough Island, where I join the six South Africans running the small mid-Atlantic Met Station, the only human presence. A much older volcano, Gough consists of a breathtaking assemblage of deep fern-filled valleys, jagged lava crags polished by winter gales, dizzying cliffs streaked with thin, wavering waterfalls, and in the mist-shrouded interior, vivid pink and green floating sphagnum bogs. Dotted over the gentle grassy slopes surrounding these bogs, well away from scrutiny, some 1500 to 2400 pairs of

34 • Spirits of the Oceans Wild

Tristan albatrosses come to nest each year, the entire world population.

Super mice

But here a bizarre tragedy has struck. Ordinary house mice that escaped from sailing ships long ago — as likewise they have on Antipodes in New Zealand — have recently developed into what scientists have named 'supermice', attaining proportions over twice their normal size. These monster rodents have acquired a bizarre predilection for burrowing through warm dry albatross nest linings during the winter and slowly devouring the hapless chicks while still alive (see Chapter 24). That such a horrible fate should befall an albatross species on one of the remotest islands on earth is a true wake-up call; teaching us how careless human actions may have dire consequences. Is it only a matter of time before the mice on Antipodes Island develop the same ghastly trick?

RIGHT Seen from the vertiginous south cliffs of Adams Island, a Gibson's (Antipodean) albatross glides over a wild, reef-studded ocean.

ABOVE Rarest of all, a pair of Amsterdam albatrosses engage in courtship on their namesake island. Photo courtesy of Scott Shaffer.

Amsterdam albatross

In the Indian Ocean over 7000 kilometres (4375 miles) to the east of Gough is another rarely frequented volcanic island, Amsterdam, this one French-owned. Here nests the rarest of all albatross species, the Critically Endangered Amsterdam albatross. Dark brown plumage, except for a white mask and pale undersides, lends it a fledgling Wanderer look. Yet with the black cutting edge to its bill resembling that of a royal albatross, its kinships are somewhat mysterious. Even more baffling is the fact that the entire species is represented by just two dozen nesting pairs, or perhaps 130 individuals worldwide.

The one thing that all these great albatross species hold in common, as indeed do all albatrosses around the world, is that the islands where they nest, with all the perils that might be encountered there, are by no means their preferred element. Coming to land — as remote as that speck may be — is a necessary risk they must take in order to breed. But their real habitat, the place where they spend the vast majority of their long lives, is in fact what we landlubbers often see as a no-man's-land: the open ocean. It is here in the immense southern reaches of our globe, where few land masses interrupt the free-ranging storms that circle the watery earth, that many albatrosses make their true home, out of sight and out of mind to most of us. And it is also here that more and more fishing fleets pursue the deep riches of the seas, bringing home cheap protein to feed the modern world and its livestock with fishmeal and other products, sometimes at the expense of the great Wanderers' survival. It is hard to project our consciousness that far, but slowly we are learning that there must be a balance between our needs and theirs — a place for both them and us.

ABOVE A Gibson's albatross makes a final approach towards its nest on Adams Island.
BELOW A courtship sequence of Gibson's albatrosses on Disappointment Island shows complex moves that begin with a duo and continue as a trio.

2
Racing Royalty
The Royals

Southern royal albatross

There can be few more exhilarating places to be than in the heart of royal albatross country. Campbell Island is roughly 650 kilometres (400 miles) south of New Zealand and the last bastion of land before the distant ice walls of the Antarctic coast, as wild a place as any on earth. High on its wind-blasted crags I sit watching dancing sunbeams play between black snow squalls about to assault the island from the south-west. All around me large clumps of snow tussock shimmer as they stream and tremble in the wind. Startlingly, the most exposed slopes are clad in fields of gaudy floral displays belonging to several species of megaherbs, or giant versions of alpine wildflowers. With rigid stalks laden in oversized purple daisies, and corrugated, ground-hugging, cabbage-like leaves up to half a metre long, *Pleurophylum speciosum* is strangely impervious to the lashing weather. In this extraordinary setting dozens of

ABOVE Their beaks soiled from nest building, mates spend much time delicately preening each other's heads and necks, Campbell Island. Photo: Mark Jones.
LEFT A southern royal albatross incubates amid fields of summer blooming megaherbs. Photo: Mark Jones.
OPPOSITE A pair of southern royal albatrosses exchange animated greetings during change-over at their nest high on Campbell Island.

Spirits of the Oceans Wild • 41

ABOVE LEFT Golden sunset rays caress a peaceful scene – a rare event on Campbell Island – as a pair of southern royals take a break from their busy courtship in the tussock grass fields.
ABOVE RIGHT In a classic gamming group, southern royals on Campbell Island gather to establish partners for seasons to come.
RIGHT Bright sun makes the snow tussock shimmer around a stolidly incubating southern royal albatross.
FAR RIGHT Dextrous flight adjustments are needed to maintain control while landing on Campbell Island. Photo: Mark Jones.

southern royal albatrosses literally drop in from the sky. Far in the distance I can see them working their way upwind, skimming over the sea, then rising over the land along the slope and over the ridge, and parachuting down with wings dipped and webbed feet splayed. Whiter still than the whitest of wandering albatrosses, and marginally larger, they cut splendid figures among the flowers, herbs and golden grasses. More stately in their manners than their wandering cousins, their gamming groups express poise as they touch bills, nod and croon to one another, from time to time throwing their heads back with an intriguing sound akin to rolling marbles. Only occasionally does their collective excitement rise to a crescendo, wings outstretched and bills pointing skywards, accompanied by shrill ecstatic wails, similar to the wanderer's. Although it is midsummer and the sun disappears for only a few hours

42 • Spirits of the Oceans Wild

each night, snow, sleet and hail are not uncommon here, but not even fierce squalls dampen their mood.

These birds are the young hopefuls garnering experience that will gradually build the trust needed to form stable pairs. Some are also failed breeders who lost a mate or egg earlier in the season and are spending the summer affirming relationships for future nesting attempts. Incubating birds dot the view, hunched quietly on their nests, which are sparsely scattered across the landscape. Near the top of the most exposed ridge, within a radius of just a few dozen paces, lie five carcasses of adult albatross picked over by scavenging skuas, eerily splayed next to their nests amid broken egg shells. To my astonishment, upon close examination I discover each shows a large haemorrhage on its skull, apparently killed by exceptionally large hailstones.

ABOVE Southern royal albatrosses carry out their courtship with poise, here overlooking the west coast of Campbell Island.

RIGHT Handing over incubation duties before heading out to sea, a female southern royal watches closely as her mate settles on the egg, which takes more than two months to hatch.

Reunion at the nest

Not far below, a female sits motionless on her nest, her eyes sunken and bill encrusted with salt crystals, looking gaunt and dehydrated but unhurt. By sheer coincidence, her resplendent mate sails in just as I approach, his plumage fresh and clean from the sea. He talks to her, nodding and groaning. She perks up but remains reluctant to leave the nest, even though she has probably sat here with no food or water for a month or more. As she eventually lifts her breast, the egg shows a small hole from where faint squeaks emanate. They both stare down, touching their bills to the egg as if incredulous that their patient wait is drawing to a close after nearly 80 days taking turns incubating. For another month or so they will continue to brood the helpless small chick, but when it is large enough they will both spend their full time at sea searching for food and making only the briefest visits to deliver increasingly large meals. During those further seven months

LEFT Late summer snow squalls strafe Campbell Island as a pair of royal albatrosses court unperturbed, with Mount Honey, the highest point of the island at 569 metres (1866 feet) and Perseverance Harbour appearing beyond.

ABOVE Reaching the apex of its see-saw dynamic flight pattern, a southern royal banks sharply before descending into the next wave trough. Photo: Mark Jones.
RIGHT A trio of northern royals confer at the edge of the densely packed Buller's albatross colony on The Forty-Fours.

until the chick is ready to leave, it is unlikely they will ever see each other again, the presence of the other visible only through the health of their chick, for it is impossible for a single parent albatross to bring up its young alone.

For five weeks I live among the southern royals of Campbell, sometimes camping on the high ground where they nest, or more often returning to the comfort of our little sailboat, *Mahalia*. Her dry cabin warmed by a small diesel heater, she offers a most welcome home-from-home tucked away in the calm waters of Perseverance Harbour. From her cockpit I never tire of watching the incoming albatross tacking against the wind. They use the lengthy inlet as a corridor slicing into the middle of the island, not keen to leave the sea any sooner than they must. About 8400 pairs are estimated to nest on Campbell Island, making up the total world population along with another 100 or so on flat Enderby Island further north in the Auckland Islands group.

ABOVE With space limited on the small vegetated plateau sometimes washed by storm swells, the nesting colony of northern royals on The Forty-Fours, tiny outliers of New Zealand's Chatham Islands, is the densest of any great albatross species.

Northern royal albatross

The northern royal, which likewise breeds only in New Zealand, is slightly smaller than its southern brethren, and its wings are noticeably darker. This species also boasts the distinction of having founded a new colony in recent times, an extremely rare event for any albatross (see Chapter 12). About 70 years ago, it established a small beachhead near the port city of Dunedin. A few birds began to nest on an open grassy slope on Taiaroa Head at the tip of the Otago Peninsula. Intensely monitored and protected, this unique mainland population has grown over the years, hatching its 500th chick as this book goes to press.

Chatham Islands stronghold

Three summers after returning from Campbell Island we sail *Mahalia* to the northernmost royal albatross nesting grounds, located in the Chatham Islands, 700 kilometres (440 miles) east of New Zealand. Here some 7000 pairs of northern royals crowd two tiny sets of rugged offshore islets, The Forty-Fours and The Sisters, privately owned by the Chatham Islands iwi (native tribes). Through the remarkable good graces of their legal keeper, Ada Hough, I am allowed the exceptional privilege of exploring these steep-sided rocks traditionally off-bounds to women.

For four blissful days I immerse myself in an albatross realm few people have set eyes upon. But the most striking feature here is the social melee into which they are thrust because of limited space. Dependent on vegetation to build their large nests, the royals restrict themselves to the higher plateaux sporting compact carpets of an attractive endemic daisy. When these plants were washed away by ferocious winter storms some years ago the birds suffered a serious setback until the vegetation recovered. Conveniently, they leave the rockier lower areas to dense throngs of much smaller Buller's albatrosses, who depart before the winter sets in.

Spirits of the Oceans Wild • 47

ABOVE Only days old, a northern royal albatross chick on The Forty-Fours is closely guarded at all times by one of its parents, protecting it from predatory skuas and extremes of weather.
RIGHT The endemic button daisy, *Leptinella featherstonii*, provides both nesting material and important shelter on The Forty-Fours.

In mid-February the islands are peppered with fresh hatchlings looking gawky with their drooping beaks and sparse white down. The colonies are a fascinating hubbub of activity as both chick care and fervent courtship go on helter-skelter at this time of year. I know that adults are making only short feeding trips, returning frequently and spending an unusual amount of time ashore. Like other great albatrosses, their slow chick-rearing process will take an entire year to complete, meaning that at any one time at least as many birds are away roaming the oceans on their year off. I am awed to consider the findings of satellite tracking studies showing that non-breeding northern royals spend much of their winter feeding in waters off the coast of southern South America, often rounding Cape Horn and foraging over the Patagonian Shelf on the Atlantic side of South America (see Chapter 17). To return, they simply resume their eastward flight, looping clear around the globe and sometimes travelling

non-stop back to their nesting grounds in just three to four weeks. The total distance covered by each bird is a breathtaking 27,000 kilometres (17,000 miles), without even accounting for the many zigzags and detours their flight path must entail.

On my last morning at The Forty-Fours, leaping ashore on a kelp-covered ledge to scramble up the cliff face and onto the summit plateau, I discover a splendid male albatross stuck in a turbid fur seal pool at the bottom of a narrow fissure. Clearly, he'd toppled off the cliff edge having missed his landing on the plateau above. He cannot escape, unable to spread his wings between the rock walls, and is paddling about frantically to avoid the playful leaps of rambunctious seal pups.

I know his predicament is a natural event and that in the purest of senses I should not interfere. But neither can I bear to let his demise pass with indifference simply because he misjudged his landing approach in the gusty wind. I scramble

down, grab his snapping beak and tuck him unceremoniously under my arm, discovering anew just what a huge bird this is. I watch him sail away majestically over the ocean after launching him from a promontory, cheering the very fact that such a phenomenal bird exists.

ABOVE Northern royals on The Sisters, north of the main Chatham Islands, court amongst button daisies which have regrown since severe storms destroyed much of the nesting habitat in the 1990s.
LEFT Although its favoured feeding areas are around the continental shelves of southern South America, a northern royal returns unerringly to the Chatham Islands to nest.

LEFT Classic subantarctic weather, marked by fast-moving fronts and sudden squalls, is normal on Campbell Island, the world capital for the southern royal albatross.
BELOW Unlike some smaller species, southern royals use well-measured gestures in courtship.

ABOVE Gaudy fields of spectacular megaherbs, *Pleurophyllum speciosum*, erupt along the windward western cliff edges of Campbell Island, near where southern royal albatrosses nest on grassy ridges and slopes.

Spirits of the Oceans Wild • 51

3
Travellers of Wind and Wave
The Black-Brows

Black-browed albatross

Imagine a dense field of white lilies stretching over an area three kilometres long and half a kilometre wide (about 375 acres). But instead of delicate blossoms you are looking at the immaculate white heads and orange beaks of tens of thousands of black-browed albatrosses. And between them, seemingly packed as tightly as apples in a fruit crate, are nearly twice as many rockhopper penguins. This is the incredible view that met my gaze 20 years ago as I rounded the peat-mudded brow of low-lying Beauchêne Island some 60 kilometres (40 miles) south of the main Falkland Islands. Back then an estimated 160,000 pairs of albatross nested amid a mind-boggling 300,000 pairs of rockhoppers, an avian metropolis of staggering proportions. To stand at Beauchêne's tussock grass margin overlooking this sprawling colony is literally to overwhelm the senses. I spent seven memorable days camped there in total seabird immersion.

Spirits of the Oceans Wild • 53

ABOVE With nesting space at a premium on Beauchêne Island, even small rock islands in a guano-stained pond are occupied: a parent prepares to feed its begging chick while another adds mud to its pedestal nest.

RIGHT Within the bustling Beauchêne Island colony, a mob of striated caracaras feast on a starved black-browed albatross chick.

Beauchêne Island life

At first I found it impossible to see detail within the mass, the sounds and smells all blending into one great carousel of life and motion. Albatross guano coated my tent and clothing, and feathers laced my food and drink. Some six weeks remained before the chicks would fledge and the colony would be deserted en masse for winter. A frenetic sense of urgency could be felt everywhere. Birds were streaming in from the ocean in a continuous flow, while others departed by making use of a communal runway devoid of nests, where they could run, wings open, to gather speed for take-off. During early mornings and evenings the noise rose to a crescendo as parents came and left, and insistent chicks begged almost incessantly for food.

Only after several mesmerising days did I begin to pick out the intertwined facets of life playing out for the individuals and their families, rather than

ABOVE Greetings are an important part of courtship for a pair of black-browed albatrosses on New Island in the Falklands, tending an empty nest which they will likely occupy during the following breeding season.

the collective mass. A returning adult would swoop low towards its nest, but each time be caught by a wind eddy and have to abort his landing at the last minute, circle and try again. Another, coming in too confidently, crashed several metres short of his target, and was immediately set upon by several neighbours and especially an irate pair of penguins who regarded the intrusion as a violation of their personal space. A pair, quite by chance, arrived at their nest within minutes of each other, this perhaps being their first home visit to coincide since they stopped brooding their chick three months earlier. Upon sighting each other they immediately lost themselves in a flurry of greetings, momentarily forgetting their frantically begging chick. Minutes later he received double rations, leaving him so replete he could barely support himself on the nest. Nearby another chick sat gaunt and panting, eyes half shut, slowly starving. The next morning life had drained out of him and a mob of striated caracara, the endemic scavenging raptors of these islands, were feasting on his meagre remains. Clearly, at least one of his parents had not returned, perhaps drowned in the highly dangerous rough and tumble pursuit of scraps in the wake of one of dozens of squid trawlers I could see plying the nearby waters.

Catastrophic decline

For an avowed albatross lover like me that week was an experience of a lifetime, and the memories remain as clear as if it had been yesterday. But in the intervening two decades much has changed on this tiny speck of land. Even though the island itself remains rigorously protected, the number of albatrosses nesting there has declined by about a third, and the rockhoppers are down to less than one-tenth of their former numbers. Incredible as that may seem, it appears this tragedy is due to the far-ranging impact of insidious human activities.

ABOVE Flying through thick fog a black-browed albatross navigates back to its nest amongst the multitudes on Beauchêne Island.
OPPOSITE Harsh climate is no deterrent for black-browed albatrosses negotiating a raging storm pounding the cliffs of New Island in the Falklands, as they return to their colony on the tussock grass slopes above.

In the case of the penguins, global warming is altering marine productivity, depleting their food resources, whereas the albatrosses die as they are attracted to high-seas fishing fleets working the waters of the Patagonian region. Some drown on the multitudes of baited longline fishhooks that are set each day from horizon to horizon to catch tuna and other pelagic (open-water) fish. Others become entangled on the cables of deep-sea trawlers while jostling among the waves to catch morsels of fish washing in the ships' wakes. Being long-lived birds who take years to mature, no albatross population can sustain such high death rates.

Still, the black-browed remains the most numerous of all albatross species. With about 10 nesting colonies dotted around the south and west of the archipelago, at last count (2005) the Falklands harboured 399,416 nesting pairs, or roughly two-thirds of the world's entire breeding population. So how can a species totalling roughly 530,000 breeding pairs be considered Endangered? The answer is that even such impressive numbers cannot conceal their catastrophic losses. According to figures compiled by Falklands Conservation, BirdLife International and the Royal Society for the Protection of Birds (RSPB), in just 10 years 38,429 pairs vanished from the Falkland colonies. This represents a decline of nine per cent, or 10.5 pairs dying every single day. For a bird that lays only one egg per year, it soon becomes evident that such a trend foretells of imminent collapse.

Good news
The good news is that momentum is gathering to bring conservation organisations and fishers together to develop ways of keeping albatrosses away from the dangerous gear trailing from the stern of fishing vessels. Mitigation techniques have been trialled and widely adopted in some regions, such as the Falkland and New Zealand fisheries

56 • Spirits of the Oceans Wild

ABOVE Large Campbell albatross chicks sit impervious to rain on volcano-like mud pedestal nests that act as tiny private islands in a peat bog liquefied by trampling.
RIGHT Only days old and still safely guarded by one of its parents at all times, a small chick begs for food with gentle squeals and light bill tapping under its father's bill.

(see Chapter 21). As a result, survival in some breeding colonies is already showing a turnaround. Yet an enormous amount of work remains to be done, especially considering the large number of fishing fleets from distant countries operating in international waters. With little control and no supervision, many of these crews have no knowledge and even less regard for the impact that their bad practices are having on birds that travel freely about the Southern Ocean and nest on some of its remotest outposts.

Black-browed albatrosses are among the more far-ranging and can often be seen in strikingly diverse surroundings. If you're cruising to Antarctica they will follow your vessel down among the islands bordering the Antarctic Peninsula, or may equally be present through the Beagle Channel and Strait of Magellan, and even deep into Chile's giant fiords. They range far up both coasts of South America and South Africa,

and every so often a juvenile strays way across the equator and turns up looking for company in the crowded gannet islands off northern Scotland (see Chapter 12).

Campbell albatross

Far away from the great concentrations of black-brows found mainly in the Atlantic and Indian Ocean sectors of the subantarctic, the Campbell albatross was once thought to belong to the same species, even though it shows some marked differences. At the very northern tip of Campbell Island, 650 kilometres (400 miles) due south of New Zealand, I visited its only known nesting area, and found to my delight that it shares many of the wonderful personality traits of its relative. Firstly, it is just as sociable, with all 23,500 pairs clustered in a few tight colonies on shelving cliff terraces. Neatly spaced nests consist of symmetrical adobe masterworks growing out of a quagmire of trampled peat. Atop the carefully crafted mud pedestals, some up to half a metre (0.6 ft) high, sit snappy chicks, their rounded tums moulded to the deep nest bowls for a perfect fit. Keenly watching my approach, they swivel in unison like little fluffy grey turrets to face me as I pass, clapping their beaks to warn me that this is close enough.

But it is the adults who grab my attention with their riveting pale irises, further enhanced by what looks like carefully applied eye shadow. Animated and curious, several take time out from their loud courtship rituals to walk over and size me up at close range. Stretching their necks and staring at me almost beak to nose, those amazing glassy eyes seem to bore through me with the intensity of their intrigued stares. Once again I feel the magic of communing with creatures whose lives are still brimming with mysteries.

ABOVE Using the cliff edge for landings and take-offs, Campbell albatrosses access one of a number of colonies tightly packed on ledges and slopes along Campbell Island's North Cape, the species' only breeding site.

ABOVE Tail flared and feet down for added stability, a Campbell albatross reaches stalling speed while prospecting for a nesting site at sunrise.
OPPOSITE A precarious landing on a crowded ledge at Campbell's North Cape draws indignant stares an instant before the collision.
BELOW The courtship routine of black-browed albatrosses involves gaping and nodding, bill touching, fake preening and tail fanning, all accompanied by high-pitched wails, yapping and low grunting.

4
Dynamic Wings and Painted Bills
Grey-headed, Yellow-nosed and Buller's

Southern Buller's albatross
Dawn on The Snares, 200 kilometres (125 miles) south-west of New Zealand, is the moment when their unique character resonates with the most powerful, otherworldly impressions. Not only are the steep granitic islets deeply indented by the incessant action of the gnawing, rumbling sea, they are also crowned by a shaggy, interwoven topknot-like forest of tree daisies, *Olearia lyallii*. The grey-green foliated hood spills over the cliffs and tumbles down steep gullies, overhanging the sea in many places.

Come the first hint of daybreak, the strangest clamouring begins to emanate from these dark thickets. The sound escalates until it overwhelms the roar of the wind, pulsating louder and louder from every valley and ridge, giving the impression that the land itself is lowing like a huge, restless beast. Only as daylight gives form to the canopy line against pink scuttling clouds are the authors of this unearthly din revealed. In a crescendo of sound and motion hundreds of thousands — nay, two million — sooty shearwaters are departing for the day. They stream out of the forest like a feathered torrent and launch themselves into the air from jutting boulders, muddy ledges and gnarly tree trunks. Mottled petrels, prions and tiny diving petrels are also departing, although they do this more surreptitiously than the multitudinous shearwaters.

Once the sun peeks over the horizon and the last of the night shift has departed, the islands return to daytime normality. Small fernbirds and even tinier all-black tomtits busy themselves picking insects along the forest margin, while a shy snipe ushers its chick into the concealing undergrowth. A bevy of Snares crested penguins, endemic to the islands, go tottering up a narrow streambed into the forest interior, screeching and squabbling as they slither and splash through dark puddles. Further up they come upon a large sleeping bull Hooker sea lion (a species also restricted to New Zealand's far

ABOVE With its colourful markings, the Atlantic yellow-nosed albatross is elegant and dainty, the two very similar yellow-nosed species being the smallest of all albatrosses. OPPOSITE An Atlantic yellow-nosed albatross begins nest building amid palm-like bog ferns, *Blechnum palmiforme*, on Nightingale Island in the Tristan da Cunha group.

ABOVE A pair of Buller's albatrosses court along the western cliff edge of North East Island in The Snares group before ducking into the forest to find a sheltered nest site. RIGHT This bird had to walk some distance to reach its nest deep in the dense, gnarly *Olearia* forest.

south) and scatter as he growls and lunges at their passage.

Even in daylight, the forest is an eerie place. The rough, wind-tortured trunks are warped and tangled. The thick canopy of broad leathery leaves, only a few metres high, keeps out most of the buffeting wind, as well as the light. The bare peaty ground, riddled with thousands of gaping petrel burrows, is slimy black and as slippery as oil. At pains to avoid collapsing petrel nests I duck, bend and wriggle over, under and in between trunks and branches twisted like the crooked arms of trolls.

Forest albatross

This Tolkienesque world seems like the most unlikely place ever to encounter an albatross colony. And yet The Snares are home to more than 8000 breeding pairs of the small and elegantly coloured Buller's albatross, the species' southernmost nesting colony. Some researchers

ABOVE The vast majority of Buller's albatross nests are tucked away out of the wind beneath the leathery-leafed canopy of tree daisies, *Olearia lyallii*, endemic to The Snares, an unusual setting for albatrosses.

distinguish it as the southern Buller's albatross, slightly larger than its more northerly relative from the Chatham Islands. Nimble in flight and agile on foot, they land on grassy ledges and rocky outcrops, then walk confidently into the shadowy forest where they build beautifully crafted mud pedestal nests tucked between leaning tree trunks. They are alert and active, if slightly less stately than their larger cousins. In the eye of the beholder this is more than compensated by the exquisite colours of their grey and white facial patterns, ashy forehead and utterly striking orange and black bill markings, which they are not shy to use to best advantage during courtship.

The southern Buller's of The Snares are also among the most thoroughly studied of all New Zealand albatrosses (see Chapter 18). Season after season the survival rate and nesting success of known pairs have been recorded and the first return of their offspring logged many years later.

The picture that has emerged demonstrates how closely linked this albatross's fate remains to fishing activities. While The Snares population benefits from scavenging behind ships, those from the Solander Islands (only a couple of hundred kilometres, or 125 miles, further north) suffer, perhaps reflecting different fishing practices on their respective feeding grounds.

Northern Buller's albatross

Far to the north and east of The Snares, in the Chatham Islands, the slightly smaller northern Buller's albatross nests alongside the northern royals on The Forty-Fours and The Sisters, tiny satellite islets well away from the main Chathams group. Although considered by some to be a separate species called the Pacific albatross, most researchers agree that the variations aren't sufficient to warrant the split. The contrast between their respective nesting habitats, however, could not be

Spirits of the Oceans Wild • 65

ABOVE LEFT At about two weeks of age an unguarded Buller's albatross chick on the Chatham Islands' Forty-Fours defends itself by rotating like a little turret, ready to project foul stomach oil towards any approaching danger.

ABOVE RIGHT A hungry chick on The Forty-Fours receives a large delivery of concentrated, semi-digested fish oil passed neatly from gullet to gullet, the rich supply being the product of many days foraging at sea by its parent.

more extreme. Unlike the sheltered depths of The Snares' forest, on The Forty-Fours not a blade of grass grows between most of the approximately 16,000 or 17,000 nests crammed within just a few hundred square metres of available terrain. Every bit of space on the narrow rocky plateau, and even the ledges along the surrounding cliffs, are literally alive with Buller's. In a neat division of territorial claims, the greener, higher ground is mainly reserved for northern royals. The segregation appears to stem from the effect of winter storms that may wash over parts of the island when the Buller's are absent, making lower ground untenable for the royals who raise their chicks throughout the winter months. Where the two species meet it is amusing to see that invariably it is the smaller, cockier Buller's who have the last word, clearing space around themselves with quick jabs and loud insults, forcing the more stoic royals to detour at a safe distance.

The Forty-Fours

I spend many long hours watching the dynamics of the colony. The density of the nest spacing seems to be determined by two factors: The availability of mud with which to construct the all-important cupped pedestals, and the distance of a jabbing beak's length to maintain a circle of personal space for each pair. This creates many a comical situation, like mud-thieving and wingtip-nipping, both of which elicit indignant screams from the aggrieved.

Everywhere fluffy chicks only a couple of weeks old are learning how to look after themselves while their parents have left them unattended during feeding trips, their biggest challenge being to duck in time to avoid being bowled over by a clumsy landing. Snug in their terracotta nest bowls, they adroitly shake off rain from their downy coats to prevent it puddling beneath them. But once toppled onto the bare ground, even though their parents may still feed them, they appear doomed to a cold

ABOVE A Buller's albatross of the northern race, slightly darker and smaller than its southern relatives, makes a final approach towards its nest on The Forty-Fours, rugged Chatham Islands outliers.
LEFT A dense colony of northern Buller's albatrosses takes up all available space, nesting on bare ground that can sometimes be washed over by storms.

ABOVE A pair of grey-headed albatrosses court tenderly while investigating a prospective nest site on Campbell Island's North Cape.

and soggy end mired in guano-rich slime. Try as they may, their weak legs and pendulous bellies prevent them climbing back aboard. As I amble through the colony, I cannot resist slipping my hand gently beneath the soft bottoms of these pathetic castaways, slowly lifting them back up. At the risk of being accused of meddling with nature, it gives me no end of pleasure to watch their bewildered looks change to contentment as they settle comfortably into their familiar nest cups.

Grey-headed albatross

Seeing a grey-headed albatross at sea, with its velvety-grey plumage and fine black and yellow beak paintwork, invariably brings a moment of excitement. It dips and veers in what seems like a slightly erratic flight path, yet performs a buoyant form of dynamic soaring with grace and confidence, reminiscent of the larger petrels. But often I find this albatross simply appears alongside, literally out of nowhere, perhaps because its darker colour tones make it less noticeable over rough seas than its great cousins the Wanderers, with whom it associates.

Questions burn in my mind: Where are you going? Where have you been? The peripatetic grey-headed can be sighted almost anywhere in the notoriously fierce southern latitudes known as the Roaring Forties, Furious Fifties and Screaming Sixties. Unlike its smaller relatives, the Buller's and yellow-nosed who tend to spend the bulk of their time comparatively close to their breeding islands, the grey-headed is the ultimate traveller. Only with the advent of lightweight, high-tech recorders has it been possible to track such extraordinary prowess in recent years. Although pairs need only seven or eight months to raise their chick, for reasons known only to them, successful parents will not return to nest again the following summer. Instead, they take an entire year off, a kind of albatross sabbatical,

during which they remain continuously at sea.

With nesting colonies on remote islands scattered around the southerly regions of each major ocean, in times past it seemed reasonable to assume that birds seen at sea belonged to those colonies nearest them. But not so. Twelve of 22 grey-heads fitted with tiny leg-mounted data loggers returned to their South Georgia nesting grounds after their 18 months' absence having happily navigated clear around the planet. Moreover, two of these 12 had actually done so twice! Equally amazing, each one revisited favoured feeding areas on each circuit, often averaging 950 kilometres (600 miles) per day in between. One male performed his entire round-the-world jaunt in just 46 days (see Chapter 13).

My first encounter with grey-heads on shore was on the Diego Ramirez Islands south of Cape Horn. Specks of land beyond the tip of the South American continent and the last vestige of the great Andean Cordillera, these wave-battered shores

ABOVE After feeding its ravenous chick which is still clamouring for more, a grey-headed albatross on Campbell Island yawns widely before taking off on another food-gathering journey.
LEFT The world's southernmost breeding albatrosses nest on the small Chilean islets of Diego Ramirez, south of storm-lashed Cape Horn.

ABOVE Atlantic yellow-nosed albatrosses — found only on Tristan da Cunha, left, and Gough Island, right — nest dispersed on rugged, rain-drenched slopes, often in sheltering vegetation.

harbour the southernmost albatross nesting colony in the world. Sharing the small tussocky islets with black-browed albatross and rockhopper penguins, the dapper grey-heads struck me as discreet and quiet. They kept to themselves away from the crowded ruckus of the neighbouring black-browed colony, choosing clean mossy swards instead of muddy valleys to nest. Here I spent my time watching pairs gingerly preening each other's facial feathers. Splendid birds indeed, they yield their secrets to science bit by tiny bit as they roam the farthest reaches of the globe.

Indian yellow-nosed albatross

The smallest of all the world's albatrosses are the little-known Atlantic and Indian yellow-nosed albatrosses, with wingspans of barely 1.8 to 2 metres (5.9–6.5 ft), and weights ranging between 1.8 and 2.9 kilograms (4–6.4 lb). These two quite similar species nest in the Atlantic and Indian oceans, respectively, and are sometimes referred to as the western and eastern yellow-nosed. The Indian species appears to be the slighter of the two by a fraction, but specific information is so scant that even this is not fully verified. The species nests only on a smattering of islands belonging to France and South Africa in the far reaches of the Indian Ocean: The French Crozet, Kerguelen, Amsterdam and Saint Paul, and the South African Prince Edward Islands. From there, when not breeding, they may range east as far as southern Australia and occasionally even New Zealand, where one lone pair has taken up residence in the Chatham Islands. For several seasons this intrepid couple has nested among the Chatham albatrosses on one of the more precipitous flanks of The Pyramid, a tiny outpost of outposts. It has been one of my long-standing dreams to behold the Indian yellow-nosed colony on the wild south-west cliffs of Amsterdam Island, a dream as yet unfulfilled.

LEFT In contrast with their cousins nesting on islands in the Atlantic Ocean, Indian yellow-nosed albatrosses form dense colonies using tussock-covered cliff ledges, here on South Africa's Prince Edward Islands. Photo courtesy of Peter Ryan.

ABOVE LEFT Spring sunshine sparkles around an Atlantic yellow-nosed albatross nesting among shoulder-high palm-like ferns in a sheltered bog covering the floor of an ancient volcanic crater on Nightingale Island.

Atlantic yellow-nosed albatross

The Atlantic yellow-nosed is very similar in appearance to its Indian cousin, except for a velvet grey head rather than a predominantly white one. Small, delicate and incredibly graceful, it also shares a predilection for nesting on some of earth's remotest islands. To reach its homeland I travelled further across the sea than I'd ever done before, following the path of the setting sun on a crayfishing vessel for 2800 kilometres (1750 miles) from South Africa. By the end of the second day I'd already enjoyed brief glimpses of my exquisite quarry, zigzagging daintily across our wake in the company of dozens of petrels large and small, as well as no less than five other albatross species: wanderer, black-browed, shy, sooty, light-mantled.

Tristan da Cunha Island

It took almost a week before the imposing citadel of a lone slumbering volcano rose from the wide open ocean roughly half way between Africa and South America. Through binoculars I could see dizzyingly thin waterfalls threading their way down among a quilt-work of near-vertical fernfields. Rounding the conical island, a small cluster of houses appeared nestled almost incongruously on a narrow verdant plain below the dark cliffs. Tristan da Cunha Island is home not only to the Atlantic yellow-nosed albatross, but also to the most isolated human community in the world, the fate of both birds and humans narrowly entwined for almost two centuries.

Landing at the settlement, a British Territory outpost, my first impression was of a genteel, orderly coastal village where rotund cows grazed in walled front yards and border collies lounged while waiting for work. Here it soon became apparent to me that the 280 staunch inhabitants have maintained close connections with traditions honed through generations of almost total isolation and, until very recently, autonomous living practices.

To one side of the tiny community an angry looking mass of fresh black lava frozen in time had threatened to overpower the houses as it wended its way down from a volcanic cone to the sea. The jagged, bristling heaps of rock left no trace of the local crayfish factory engulfed by a sudden eruption which had caused the island to be temporarily evacuated 45 years ago. At the back, the village green abutted a towering dark wall of older volcanic layers rising sheer some 750 metres (2460 ft) and disappearing into thick, swirling clouds. This heavy shroud of near-constant drizzle concealed the somewhat gentler slopes above, known to the locals as 'the base' (referring to the base of the central volcano) grazed by free-ranging sheep.

Never in my wildest dreams could I have imagined a more unexpected setting for an albatross. Heavy mists and occasional snow flurries lent mystery to a landscape gouged by sheer, gaping gulches and clad in a diminutive forest of miniature tree ferns. Shaped

ABOVE Though nesting further north than many southern albatross species, the damp fern-clad valleys on Gough Island, where this yellow-nosed albatross is investigating an old abandoned nest, are subantarctic in climate.
FAR RIGHT Nests are thinly scattered high on the misty slopes of Tristan da Cunha's active volcano.

Spirits of the Oceans Wild • 73

ABOVE A pair of Atlantic yellow-nosed albatrosses pause briefly to inspect a steep, sodden valley on Gough Island, where nests are ensconced in lush vegetation.

like tiny, squat palm trees standing rigid against the punishing elements, the ferns provide shelter for the thinly scattered albatross nests tucked between their dark mossy trunks. The contrast between the sombre, sodden setting and these immaculate birds was nothing short of extreme. With their soulful eyes and satin-soft plumage, their gold and peach-pink painted beaks and airbrush make-up, they were the embodiment of pure albatross elegance. It is mainly upon these birds that the local community depended for sustenance in times past (see Chapter 20).

Nightingale Island

On nearby Nightingale Island, where neither the habitat has been altered by livestock nor the albatrosses ever harvested, the colonies are denser and livelier. The sun also shines more often as the smaller island does not create its own cloud cap. In many places the yellow-nosed nest in thick stands of bamboo-like tussock grass some two to three metres (6.5–10 ft) high, or in sphagnum bogs, building nests of moss resembling tiny volcano-shaped islands among the water-loving palm-ferns.

Much further south, on uninhabited Gough Island, where the largest numbers breed, colonies fill wide valleys shimmering with emerald-green water ferns, and gamming groups gather daily on selected hilltops which they use as dance parlours. To me their realm is one of sparkling light and dancing colour, stark contrasts and infinite beauty, truly a world apart.

LEFT A courting couple refurbish an old nest with mud and grass in a loose colony scattered amongst deep ferns and clumps of squat island trees on Gough Island.

RIGHT A nesting Atlantic yellow-nosed albatross on Gough Island shelters amongst *Phylica arborea* trees, the only species growing naturally in the Tristan da Cunha group, known commonly as the island tree.

FAR RIGHT A group of non-breeding sub-adults hang around the periphery of the colony on Nightingale Island, courting where landing and take-off is easy.

BELOW Like others of their tribe, the courtship sequence of Buller's albatrosses of The Snares involves many distinctive postures which sometimes — but not always — concludes with mating.

76 • Spirits of the Oceans Wild

5
Down Under Specials
The Shy Tribe

Shy albatross
Why should an albatross be given the name 'shy'? This puzzled me for a long time, until I learned that its provenance apparently stemmed from the fact that it did not tend to follow ships like most others do. Unfortunately, in modern times these rather unusual looking albatross have lost their temerity and, to their detriment, joined the throngs tempted by fish scraps and lethal baited hooks trailing behind high seas fishing fleets.

What was once regarded as a single species found around the southern parts of Australia and New Zealand has now been recognised as four distinct types. Hence the 'shy' moniker now refers only to the Australian species, *Thalassarche cauta*, sometimes also called Tasmanian shy. Its nesting range is limited to three small islands, Pedra Branca, Mewstone and Albatross, not far offshore of Tasmania, where a total of about 12,000 pairs nest.

White-capped albatross
The seafarer who coined the name Disappointment Island (the westernmost of New Zealand's subantarctic Auckland Islands) must have found himself in a pretty distressed state. Indeed, one of the most epic survival stories is that of the 16 battered men who scrambled up the storm-lashed cliffs of this forsaken island to escape the sinking barque *Dundonald* in 1907. With only the clothes on their backs they kept themselves alive for three months eating seals, albatrosses and megaherbs before seeking rescue in a sealskin rowboat.

What this desperate crew probably didn't realise amid their ordeal is that they had in fact been shipwrecked on one of the most albatross-rich islands in the whole of the Southern Pacific sector. With around 110,000 pairs of white-capped albatrosses, Disappointment Island is by far the stronghold of the species and as a result has become one of the most highly protected seabird

ABOVE Uttering a siren-like wail while gaping widely, a courting Chatham albatross holds the full attention of its prospective mate on The Pyramid.
OPPOSITE From the summit of Proclamation Island a pair of Salvin's albatrosses overlook the ragged granite outcrops that make up the Bounty Islands, the species' primary nesting domain.

islands in the world. An additional 3000 pairs nest on nearby main Auckland Island and a few dozen more on Adams Island just to the south.

Catching the brunt of the westerly weather, weeks can pass without a chance to approach Disappointment Island. During one of my own sailing expeditions, after weeks of waiting, I had the opportunity to accompany a science party there during a narrow window of calm weather. When the lucky day finally came, I could hear the albatrosses almost before I could see them. Far in the distance the dark green slopes were dotted with innumerable white flecks, each representing a bird on a nest, while on the wind drifted a strange cacophony of high-pitched, almost tweeting sounds from the throngs busy greeting and courting. On land the going was surprisingly difficult through waist-high vegetation fertilised not only by the huge albatross population but also by many more thousands of nocturnal petrels

whose burrows were almost impossible not to crush. Everywhere grave-looking albatross parents stood over snow-white chicks, fussing, preening or greeting one another with those strange piercing calls. Pairs performed their rather different version of the now-familiar albatross dance, their pale ashy plumage and graceful gestures, bowing and curtsying with rigidly arched necks, reminding me strangely of porcelain dressage horses. Meanwhile, swarms more swirled in from the sea, giving the throbbing island a life all of its own, pristine in every respect, its ecosystem intact — nothing added, nothing taken — even the briefest human incursion like ours being a rarity.

Later I camped in bitter cold wind and rain on the cliff edge of South West Cape on Auckland Island, where the nesting white-capped are not nearly so lucky. Here they are beset by feral pigs left behind from a failed attempt to colonise the island in the nineteenth century. In horror, I

ABOVE LEFT Rain droplets beading on their immaculate plumage, a pair of courting white-capped albatrosses at Auckland Island's South West Cape are totally engrossed.
ABOVE White-capped albatrosses flying by the high cliffs of Auckland Island towards their nests on Disappointment Island may be returning from as far as South African waters.
FAR LEFT Storm-lashed Disappointment Island is the last bastion of land west of Auckland Island.
LEFT Nearly the entire world population of white-capped albatrosses breed on Disappointment Island.

watched helplessly as a bristly black sow yanked a hapless chick from its nest, its desperate bill clapping and defensive vomiting no deterrent to the marauding beast. Within instants nothing more remained of the chick's plump little body and snow-white down than muddy shreds of flesh gulped down in ravenous mouthfuls.

When I emerged from my tent next morning I found one of the chick's parents had returned to deliver food. Standing alone on the empty nest, feathers ruffed, the bereaved parent stared again and again at the space between its splayed webbed feet where its young should have been, dipping its head with exaggeratedly slow gestures. Finally, its mind made up, it gulped hard to push back the gulletful of food it had been unable to deliver, and opened its slender wings into the wind. I watched the majestic bird sail away until it vanished in the far distance, drawing a straight line towards the familiar sanctity of the open ocean, its real home.

Salvin's albatross

Seeking out another relative of the shy, the Salvin's albatross on its nesting grounds — one of the world's least accessible — was a long-standing dream. To get there we devised our own private expedition, including three volunteers, combining science and photography under the auspices of New Zealand's Department of Conservation. Here I recount impressions from my field notes:

Fog descends and the breeze drops to a zephyr just as land looms ahead through the mist. No colour at all greets the eye on the pale granite outcrops, liberally splattered with guano. Yet an astounding amount of life clings to these tiniest of salt-sprayed shards of land. Mahalia, *our 13 m [42.6 ft] steel cutter, approaches these enigmatic islands as an eerie calm falls upon us.*

The Bounty Islands — discovered in 1788 by Captain William Bligh and named after his ship, the HMS Bounty *of mutineering renown — are the strangest and least visited spot in the New Zealand subantarctic region. A mere handful of weather-beaten rocks crammed cliff edge to wave-lashed cliff edge with seabirds and fur seals, they are the epicentre of the Salvin's world.*

Probing the seafloor by sonar, Mahalia *pulls up close under the dark monolithic cliffs, but the bottom remains shy, well over 30 m [98 ft] deep within an arm's reach of the wall. Still, this is the best anchorage to be found, and the exceptionally calm weather today is an un-hoped-for chance to set up camp on Proclamation Island. This is only the third time humans will live ashore on the Bounties since the last sealing gang worked here in 1881.*

The air is alive with the bellows and shrieks of fur seals and seabirds as carousels of albatross reel overhead and rafts of erect-crested penguins make the sea surface boil. Once ashore, every

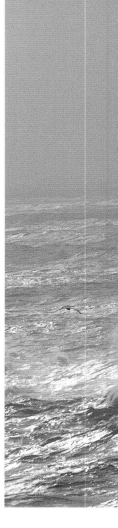

ABOVE Facing into a wet northerly gale, an incubating Salvin's albatross lets driving rain roll off of its waterproof plumage, keeping the nest bowl dry.

RIGHT With no shelter from the lashing storm, our camp is precariously perched among nesting birds on Proclamation Island.

granite ledge and furrow is occupied by a resolute-looking penguin hunkered down on its egg, roguish golden crest bristling and flippers flailing, squawking indignantly at our awkward scramble up the cliff. By contrast, the albatrosses are all poise and dignity, perched upon scanty crescent-shaped mud nests. They stare down their oddly drooping bills with dark, pensive eyes accentuated by ashy white brows and a brush of charcoal eye-shadow. Thirty metres [98 ft] above sea level, we pitch camp on a narrow platform of awkwardly sloping, slippery rock, sharing the vaguely flat slab with three albatross and six penguin nests only a beak's length away from the tent fly.

By morning the whole world has changed. An angry north-easterly storm is tearing at our wildly flapping tents. Mahalia *has found questionable refuge in a rocky cove beyond our sight as huge waves pound the cliffs where we landed, driving*

ABOVE Storm waves pound the granite walls of Lion Island in the Bounties while returning albatrosses ride the gusts unperturbed.

spume and rain over the island. Dubious-looking greenish-brown guano slime oozes down the slope, through camp and under the tent floor. Yet the birds around us are serene, leaning into the wind and letting the water run off their beaks and feathers. Quickly I realise that this is life as usual on the Bounty Islands. There is utter peace in such wildness.

The next day the storm abated and the sun returned. Enveloped by albatrosses on all sides, life took on that perfect, raw, basic tempo that can only ever be found in places truly untrodden and untamed. With the island only 300 metres (985 ft) or so across, I dedicated myself to seeking out individual details among the seething masses. A pair fussed intently over a crushed egg, seemingly unwilling to admit failure for the season. Meanwhile the 'camp albatross' nesting directly by the tent door hatched a tiny, innocent-looking chick. Mum became totally accustomed to our presence, tapping me gently on the shoulder one evening when I inadvertently sat eating my dinner with my back in her face. But the next day Dad arrived to take over the watch and I found out to my detriment that he had very different ideas about personal space, bloodying my hand as I reached for some gear.

One pair on the summit ridge intrigued me particularly. Both were wearing metal bands — numbers O-14211 and O-14217 — while almost no others around them did. Later I tracked down their history: Both banded as chicks in 1985, O-14217 had not been seen since. They would now be 20 years old, and it seemed quite extraordinary to me that they had chosen each other among the population of unbanded birds.

With the passing days I became more aware of developments in the island's biological clock. Chicks began to hatch everywhere, delicate fluff balls with oversized drooping beaks and beady black

Spirits of the Oceans Wild • 85

eyes, over which parents doted as if in disbelief that such a wonder could appear out of the egg they had steadfastly sat on for over two months.

On the fourth day of our stay, a tremendous influx of non-breeding albatross converged on the island like an aerial river of birds flowing in from the east. Competition for space became dramatic, vicious fights erupting as arriving birds tried to secure new nesting spaces. Locked in battles, they tumbled among the penguins, who in turn responded with volleys of jabs and flipper beatings. Fur seals trying to reach favoured sleeping spots ran a gauntlet of snapping beaks. Yet high on a promontory I saw one young bull irately grab an unwary albatross and toss him clean over the cliff edge. Perhaps to counter the constant roar of wind and wave, the sheer volume of all these birds' voices far exceeded their brethren elsewhere. Whether courting or defending their turf, the high-decibel cacophony had my eardrums throbbing.

In the total absence of vegetation there is no soil

at all on the Bounty Islands, making nest-building mortar a precious commodity. Instead, each albatross used compacted 'mud' consisting of raw organic debris — spilt food, excreta, bones, moulted feathers and fur seal hair — carefully cemented to the granite bedrock. Stealing mud or even quarrying a neighbour's nest was invariably cause for a nasty fight, and many nests amounted to no more than crescent-shaped rings barely preventing eggs or chicks from rolling down the slope.

Information about the Salvin's albatross remains scant. Little is known about how and where they feed, and counts are so few that data is lacking to gauge the health of the population. We were lucky to have Jacinda Amey with us, an expert and tireless conservation field worker who had conducted one of only two previous bird counts here. Thus our expedition was able to make a census of every nest on Proclamation Island, repeating the methods she had used seven years previously. The area was divided into eight blocks and a dot of orange water-soluble paint dabbed by each nest. With 2634 nests counted, the total figure fell 412 short of the previous results, a nearly 14 per cent drop. Although the Bounty Islands revel in their pristine status under rigorous protection from New Zealand's Department of Conservation, more effort may yet be needed to ensure the survival of the Salvin's albatross on the high seas.

Sailing home from the Bounties, just as we were about to leave albatross waters behind, *Mahalia* ran into another storm. At daybreak on our next to last day of the voyage a wall-like rogue wave slammed her broadside. Over she went, mast under water and sails shredded — her first knockdown. Springing back up almost immediately, enough water made it down below that in just 20 seconds all our electronic gear was dead. Humbled though not beaten, we spent two months repairing and replacing, preparing for our next albatross-seeking voyage.

ABOVE A day-old chick emerges briefly from the warmth of its parent's brood patch to receive its first meal in tiny beakfuls.
ABOVE LEFT With the sun about to set beyond Spider Island in the Bounties, more arrivals crowd the already tightly packed colony, where Salvin's albatrosses vie for space with erect-crested penguins and in some places also fur seals.
FAR LEFT Competition for space on the Bounty Islands is intense, causing many fights and confrontations.

88 • Spirits of the Oceans Wild

LEFT With hatching in full progress on the Bounty Islands, great masses of adult Salvin's albatrosses return from the sea.
FAR LEFT ABOVE A Salvin's albatross swoops in for a high-speed landing at the Bounty Islands.
FAR LEFT BELOW High on the wind-polished granite slabs of Proclamation Island a pair of Salvin's albatrosses dance in the rain.

ABOVE LEFT A pair of Chatham albatrosses attend their carefully crafted nest built of guano and feathers deep inside The Pyramid's large cave.

ABOVE RIGHT Tail fanning and mutual gaping marks the high point of a long and involved courtship session for a pair of Chatham albatrosses on The Pyramid.

RIGHT Non-breeders gather to socialise on a flat ledge near the summit of The Pyramid.

Chatham albatross

On our next foray we were heading for an even more challenging rock, as well as the Holy Grail of the albatross world: The Pyramid, an aptly named pinnacle jutting from the seas south of the Chatham Islands and home to the Critically Endangered Chatham albatross. The species' entire world population — all 5300 or so pairs — nests on this single forbidding volcanic pillar (see Chapter 19).

Once again, there were adrenaline-pumping moments leaping ashore from our bouncing inflatable, hanging onto kelp and hauling gear by rope through frothing waves and up the cliff. But the rewards were extraordinary. The island, at 170 metres (557 ft) high, is nearly as tall as it is wide, the solid, black core of an ancient volcano, its sides so steep they are smooth and slippery, mostly clear of any loose matter. Yet from top to bottom nimble

90 • Spirits of the Oceans Wild

albatrosses plaster their nests to rock ledges, taking advantage of clumps of pink-flowered ice plants and other lush herbs growing out of tiny fissures. Theirs is a near-vertical world, where worsening weather and climate change appear to be causing vegetation loss that is jeopardising the birds' tenuous footholds.

Near the base of this monolith the only campsite is swept clean of rubble by the occasional wave-wash, the bare rock so polished that fishhooks jammed in fissures were needed to secure tent guys. I soon found the only way I could feel safe moving about on these frightening slopes was in bare feet to provide added grip.

Everything on this citadel came in megadoses of powerful impressions. Sidling around the vertiginous face led to an enormous cave, its dark gaping maw revealing a dense village of albatross nests that, never exposed to rain, have grown over generations into bizarre,

ABOVE Advertising for courtship, a Chatham albatross standing on a freshly built nest chatters and bows, briefly exposing its gape line of bright orange skin normally hidden under cheek feathers.
LEFT A plump chick sits atop a perfect nest which is carefully refurbished every year.

ABOVE With the islands after which it is named visible in the distance, a Chatham albatross approaches land only at The Pyramid.
RIGHT In the rainless shelter of The Pyramid's giant cave, albatross nests form tall pedestals.

metre-tall (3 ft), amphora-like structures. Far below, huge swells thundered around the unyielding basalt ramparts, rolling up from the far Southern Ocean like small tsunamis. Swirling fogbanks spawned by cold sea currents hugged the overhanging cornices, while far above, the summit ledges were often bathed in warm sunshine, with fogbows traced in cotton-wool cloud wisps. This was a magical, utterly inhuman, awe-inspiring place, a fitting home for one of the world's most endangered albatrosses.

LEFT Small clumps of vegetation growing from fissures form critical bases enabling Chatham albatrosses to build their nests on The Pyramid's steep rock face.

ABOVE Tail fanning and wide gaping are important features of Chatham albatross courtship.
RIGHT From the 170-metre (557-ft) high summit of The Pyramid, a Chatham albatross surveys the cool surrounding ocean as the sun paints a fogbow in the mist below.
BELOW The shy albatross tribe, consisting of four species, all share similar courtship rituals, such as these performed by Chatham albatrosses on The Pyramid.

94 • Spirits of the Oceans Wild

6
Enigmatic Elegance
The Sooties

Light-mantled albatross

Without a doubt, the most mysterious as well as the most un-albatross-like albatross is the light-mantled, also called the light-mantled sooty. It is distinctly one of the least colourful of all albatross, yet few who have seen it have failed to fall under its quasi-mythical spell. Its heart-stopping grace and understated beauty, and the way it imprints upon the subconscious the very essence of the wildest of wild islands, all seem irresistible. Not only does it nest on just nine of the most isolated islands nearest the Antarctic Convergence (four in the New Zealand/Australia side of the Pacific, one in the South Atlantic and four on the Indian Ocean sector), but it does so almost secretly, ensconced in shady gullies and mossy ledges along the most wind-blasted cliffs and mist-shrouded crags. Here its eerie 'Pee-oow' cries, sounding vaguely like a distant mix between a peacock's crow and a sailor's 'Aaaa-hoy', can be heard echoing down valleys or drifting on the wind.

Again and again, the courting bird throws its head far back as it calls to those passing far below, a truly haunting sound indeed.

For my fondest recollections of time spent in the light-mantled albatross's realm I must turn to my notes of Campbell Island during a three-and-a-half-month sailing trip, *Mahalia*'s longest subantarctic sojourn:

In spite of the ferocious weather, these last few days we attempted to explore the south coast, making several trips as far as semi-sheltered waters would permit. This brought us to an intriguing bay called Monument Harbour. And what a jaw-dropping, words-defy-description site it is! Having not heard about this place from anyone previously boosted our sense of discovery. Landforms of unimaginably sheer contours — spires, stacks, needles and overhanging crags — receive the undamped

ABOVE Eyes wide, a pair of courting sooties touch bills lightly.
OPPOSITE The eerie courtship call of a sooty albatross rings hauntingly across a misty, fern-filled canyon high in the rugged interior of Gough Island.

Spirits of the Oceans Wild • 97

frenzy of the Southern Ocean in all its raw fury. Yet, surprisingly, there is a small sheltered nook with a perfect dinghy landing site, even as the wind overhead careens and carves, buffets and blasts in volleys gusting at 30–40 knots and more. Elephant seals, sea lions, rockhopper penguins and Campbell endemic shags enliven the shoreline with their activity.

The anchorage is tenuous but comfortable, almost improbable. Most enchanting of all, every crag, every high tussock bluff, every punishing wind swirl and eddy is frequented by the most mysterious, most graceful, most masterly of all albatrosses, the light-mantled. Humble in its mocha-brown cloak, yet with exquisitely subtle white highlights around dark, mysterious eyes and a fine pale-blue line traced along the sides of its polished black beak, it is both a surprise and a delight to encounter at close quarters.

They congregate here by the score, calling

FAR LEFT The aerial *pas de deux* of courting light-mantled albatrosses, performed along the flanks of Mount Honey near Campbell Island's summit, is a mesmerising sight.
LEFT Graceful elegance defines a light-mantled albatross gliding along the cliff edge of Enderby Island in the Auckland Islands group.
BELOW FAR LEFT A magical wild place where volcanic grandeur meets tumultuous seas, Monument Harbour on the south coast of Campbell Island is the light-mantled albatross's perfect realm.
BELOW The 'pee-oow' call of a light-mantled albatross breaks the stillness of fog-shrouded Cumberland Bay on South Georgia.

their strange, haunting courtship crowings and landing on teetering cliff faces to feed plump fluffy chicks, secreted on tussock ledges amid dangling ferns. Looking out at their expansive world through astonished-looking eyes accentuated by large white goggle-like face patterns, the chicks' cuteness factor of 10/10 will only ever be known to passing skuas in this forgotten corner of our planet. Kelp-wrapped shores, spume-raising williwaws, stratified red and brown volcanic cliffs, fern-filled gullies, wind-rippled tussock fields, and above all the sound of the albatross proclaiming their love on the wind are all memories that will stay with me forever.

If the light-mantled albatross's call is deeply moving to hear, its courtship flight is just as enthralling to see. The most agile of its tribe, capable of riding the most vicious downdraughts, it easily circumvents limited dancing space on high nesting ledges by taking to the air to perform a sumptuous aerial ballet instead.

ABOVE The spectacular megaherb *Anisotome latifolia*, blooming in early summer, conceals a peacefully incubating light-mantled albatross high above the kelp-filled cove of Monument Harbour, Campbell Island.

With dagger-blade wings and pointed tails held taut, would-be pairs engage in breathtaking tandem flight, careening this way and that only a beak's length apart with perfect feather-flinching control. Tracing graceful curves above crashing surf or broad valleys, they follow one another in such absolute synchrony as to appear mirrored in flight.

The light-mantled albatross's life cycle is no less mysterious, and its habit of nesting thinly scattered over wide areas of precipitous terrain makes its study exceptionally difficult. What is known is that it raises, at best, only one chick every other year, whereas on some islands such as Macquarie, as many as four years may pass between nesting attempts, making it one of the slowest breeders of any bird. It also ranges much further south than any other albatross species, down to the very fringes of the Antarctic pack ice. Its graceful dark outline skimming over sparkling, half-frozen seas is a truly unforgettable sight.

Sooty albatross

Before I came to Gough Island in the South Atlantic never had I met an albatross and its island simultaneously. Landing here required clambering from a small fishing tender into a dangling rope basket and being craned up the 30-metre (100-ft) cliff by the small team manning the South African meteorological station, the only human presence on the island. Part way up the cliff, to my utter surprise, I suddenly came eye to eye with a sooty albatross serenely incubating its egg on a peaty ledge. I barely recovered from my amazement when I met the six smiling faces on the concrete pad above. The beauty of the bird — velvet brown plumage, lace-thin white eye crescents and mustard pencil-line along the sides of its ebony bill — went on dancing in my eyes.

The sooty has a more northerly, as well as decidedly more restricted, range than its light-mantled cousin, and Gough is one of its few strongholds. It also remains one of the least understood

LEFT Surrounded by wild celery, one of the native megaherbs of Campbell Island, a light-mantled chick sits in a straw-lined nest at Monument Harbour. Photo: Mark Jones.

albatross. Like its cousin, it does not usually breed in consecutive years, even though raising its single chick takes only seven months. Very little is known about what it feeds on and especially how and where it forages. It is almost never seen approaching fishing vessels and yet most (or possibly all) of its colonies are in serious decline for unknown reasons. Even its population numbers have eluded tally because of the extremely difficult terrain it chooses to nest in, where it blends perfectly with its surroundings. Cryptic in both looks and habits, these birds personify all the mysteries of earth's extraordinarily wild places.

    I soon discovered that several small colonies nest in clusters of about half a dozen just below the station, blending into dark volcanic crannies overhanging the sea. While incubating birds sit comfortably on their crater nests, with rain sprinkling diamonds on their velvet heads, others are busy prospecting nest sites for seasons to come. Their 'pee-oow' calls, similar to the light-mantled, ring out from hidden ledges

ABOVE Tails fanned and wide-eyed, a pair of sooty albatrosses on Gough Island court silently on a misty cliff ledge.

RIGHT (both) The piercing call of a sky-pointing sooty albatross carries far above the noise of the surf below.

 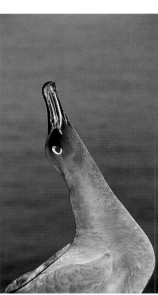

102 • Spirits of the Oceans Wild

like clear bugles over the roar and crash of the surf below, attracting a steady stream of potential mates circling low in the mist and jiggling through the squalls with ease.

Gough Island

Much higher up, others were scribing their flight path across the lowering sky towards the far interior of the island, their slim dark silhouettes outlined neatly against the scuttling clouds. Gough is an island of stark visual contrasts, and likewise are its three resident albatrosses. Many of the sea cliffs consist of chocolate-cake layers of reddish, greyish and blackish volcanic scoria and basaltic lava, usually fringed with pale wispy tussock. Yet the slopes above glow in shockingly vivid emerald green carpets of water-ferns, bristling with the neat palm-like shapes of rigid bog ferns (the same diminutive tree ferns as on Tristan) growing almost as high as a person in places. Tea-coloured streams

ABOVE Poised on a nesting ledge, a sooty albatross watches attentively for potential mates prospecting on the wing.
LEFT A courting couple veer sharply high over the verdant valleys of Gough Island.

Spirits of the Oceans Wild • 103

ABOVE Surf pounds the base of overhanging cliffs where a sooty albatross displays on a tussock grass stool.

and waterfalls cascading down numerous valleys are flanked by lichen-clad, wind-twisted copses of the island's only tree species, *Phylica arborea*, its dense grey-green foliage looking almost juniper-like. These garden-like vales and slopes are the realm of the colourful yellow-nosed albatross, as well as innumerable nocturnal petrels riddling the ground with their burrows, from tiny diving petrels to greater shearwaters.

Exploring the higher ground, the scene changes dramatically as the lush vegetation is replaced by waterlogged bogs that easily swallow your boots up to the knee. The colour scheme here changes completely, alternating between broad swatches of olive-green cushion plants, dark emerald reed patches, coffee-coloured peat bogs and quilts of floating sphagnum moss ranging from Day-Glo green to strawberry-tint. This variegated palette, riddled with dark pools and meandering streams, makes only fleeting appearances between rolling banks of ground-hugging clouds. Bands of predatory skuas gather here to digest their nightly banquets of hapless prions, and large chicks of the rare Tristan albatross punctuate the less sodden grassy slopes.

Higher still, wind-polished basaltic dykes, plugs and crags disappear bleakly into the self-generating mists that almost continuously form over the island. Only ground-hugging mosses, lichens, tufty grasses and sprawling heather-like diddledee cling to the weather-buffeted ground. This seems a most unexpected place to meet the island's only land bird, the long-legged, greenish and beige endemic bunting, happily picking tiny orange drupes and quasi-invisible seeds. And here too the sooty albatross reappears, sailing through cloud banks and swooping down into gaping canyons towards ledges accessible only thanks to its superior flying skills.

Day after day I search and probe this wild country, occasionally reminded of the two young men who died of exposure, after becoming disoriented by

104 • Spirits of the Oceans Wild

ABOVE A sooty albatross incubates in the shelter of bamboo-like tussock grass along the eastern cliffs of Gough Island.
LEFT A sooty albatross pauses before take-off on a high volcanic ridge.

Spirits of the Oceans Wild • 105

ABOVE In a rare social moment, a group of sooty albatrosses land briefly to court on a mossy ledge overlooking their nesting valley on Gough Island.

the mists. I scramble over wind-blasted ridges, down slippery faces, through dark clefts and into breathtaking canyons in search of these mythical birds. Each time I come upon a small group under a dripping mossy lip or in the shelter of dark ferns, my surprise and excitement is renewed. Sometimes they are curious, even nibbling gently at my sleeve, though more often they remain completely unperturbed by my presence, dozing contentedly on their nests without a care.

The going is rough at best, and sometimes exhausting to the point of questioning my sanity. Occasionally, when the wind is just right for their aerial acrobatics, I spend hours in fascination watching them perform their exquisite mid-air *pas de deux*. The soaring couple emulate each other's every move, their flight describing an ethereal — perhaps even spiritual — bond.

Then one day, as rain pummels the high cols, I finally discovered the object of my imaginary quest: a splendid tiered canyon where clear waterfalls tumble down a giant volcanic-scale natural staircase acting as a perfect backdrop to the sooty's wondrous theatre. Dozens of the mythical birds sweep in and out of gaps between lacy cloud banks, some in duo flight formation, others lured by the beckoning calls of those perched on unseen balconies far below.

Digging fingers and toes into soft mossy walls, I slither down into these plunging, mist-shrouded depths until I draw level with the singing cascades and hushed fern-world where courting sooties land, flicking their beaks and nodding to one another while they select their ideal building sites for new nests. Suddenly finding myself at the heart of their secret sanctuary, and so as to anchor the memories of these magic moments forever, I named this place simply 'Sooty Canyon'. I don't think in another lifetime could I ever tire of such soul-lifting grace and beauty as I discovered on that day on Gough Island.

ABOVE Wind-driven mist veils deep canyons on Gough Island obscuring the dark, secretive world where sooties nest.
LEFT (both) Blending into the perpetually rain-drenched landscape, sooty albatrosses court and nest on secluded ledges, often far from the sea.

Spirits of the Oceans Wild • 107

LEFT High over Campbell Island a light-mantled albatross makes an inspection pass along a broken cliff face in search of a nesting site. ABOVE As with the sooty, light-mantled albatross courtship alternates between tandem flights and prospecting for a nest site on a high ledge, with silent bowing, snapping, tail fanning and stamping of feet.

Spirits of the Oceans Wild • 109

7
North Pacific Survivors
The Northern Albatrosses

Laysan albatross

For 2000 kilometres (1242 miles) we seem suspended in a blue void, the entire world composed of varying shades of blue: cobalt blue above, aquamarine below, plus a few strands of greyish-blue clouds hanging in fluffy tiers between the two.

Travelling to albatross islands invariably means venturing far beyond the horizon, but never have I felt the immensity of an ocean more powerfully than flying out of Honolulu roughly in the direction of Japan, into the heart of the North Pacific. The hours pass. Just as the setting sun dips towards the western horizon, we begin our descent. Banking left, another blue hue appears beneath our wing, a circular smudge of vivid turquoise, the unmistakable footprint of an atoll.

Formed of nothing more than sand and coral rubble, Midway Atoll seems about as unsubstantial as land could ever be. A ring of white breakers traces the barrier of live coral protecting two wisps of islands less than six square kilometres (2.3 sq miles) and barely protruding from the sea.

Daylight is all but gone by the time the 737 sets down on the runway. Taxiing towards the glaring lights of the small terminal, the plane veers sharply to detour around an obstacle on the tarmac. And there, facing the big machine bolt upright, is my very first Laysan albatross, heralding another million or so more that I'll meet in the next few days. That a commercial aircraft should sidestep a single albatross on an island where some 437,000 pairs nest each year denotes a sea change in human attitudes. Now strictly protected, these birds once suffered repeated devastation on the most appalling scale.

First, Japanese feather hunters invaded the North-western Hawaiian Chain in the late 1800s and slaughtered albatrosses literally by the hundreds of thousands. Despite strong government protest, the

ABOVE A pair of young adults rest peacefully during the warmer time of day. The peach colour of their bills intensifies with age.
OPPOSITE Under tradewinds-skies, a pair of Laysan albatrosses court enthusiastically on Sand Island, Midway Atoll, near the tip of the North-western Hawaiian chain.

ABOVE The grassy beach berm on Sand Island serves as a communal landing strip where Laysan albatrosses come and go without infringing on nesting space.
RIGHT Nesting colonies occupy every available space on the low sandy atoll.

bird pirates worked through entire nesting colonies, stripping only the breast feathers and wings, and leaving the carcasses piled high in putrefying mounds. Their feathers went to stuff beds and adorn trendy ladies' coiffures around the world. On some islands cartloads of eggs were also hauled away, the albumen used in the budding photography trade. Ironically, a spate of guano mining in early 1900, while no doubt also causing disturbance to the birds, actually brought respite by impeding the inroads of the feather hunters. Likewise, when Midway became a staging post for the first Transpacific Cable in 1903, disturbance by construction crews and their pets was offset by the fact that their presence shielded the albatrosses from the annihilation wrought on other remote islands.

Soon the Second World War brought another wave of destruction as the island was turned into a major US naval base. Notwithstanding protection orders, much of the Laysan's largest breeding

ABOVE Tip-toeing to full height, a pair sky-points in unison on what used to be Midway's military airfield.

colony was laid waste to make room for the military installations. By the time construction was completed the little atoll had been entirely transfigured to accommodate several runways and many dozens of buildings, hangars and a giant fuel farm. Bristling antennas snared yet more albatrosses in their aerial cables, while tens of thousands of others were systematically exterminated over the following years in futile attempts to reduce the danger posed to aircraft by mid-air collisions.

Today, at long last, the albatrosses and other seabirds have reclaimed their island uncontested as a National Wildlife Refuge under US Fish and Wildlife Service jurisdiction, and Midway is their largest nesting colony. The solution to air strikes, it turned out, is quite simply to limit aircraft operations to nighttime when few albatross are airborne, hence our evening arrival on the commercial Aloha Airlines flight from Honolulu.

The military personnel are now gone, and so are the rats that once plagued humans and birds alike. Most of the buildings have been bulldozed into an odd, sand-covered mound which stands out from the island's otherwise flat contour. Difficulties with invasive introduced plants and trees remain, though management programmes are gradually addressing some of these issues. With its new facelift, my impressions of Midway are as close to a Pacific island paradise as I could ever imagine, even without the classic palm trees.

Winter days are cool and breezy, and flocks of ethereal white terns flutter like butterflies among the stands of introduced casuarina trees, their eggs delicately balanced on bare branches. By contrast, the black noddies nearby build solid nests of twigs.

On the beach endangered monk seals sleep without a care in the world, and even huge green sea turtles haul out to laze on the sun-drenched sand, something they do only where there is zero

ABOVE Arriving from the far reaches of the northern Pacific, Laysan albatrosses use the beach edge for lift as they round Sand Island on the final approach to their nest sites.
RIGHT A Laysan albatross banks low over the vibrant tropical lagoon.

disturbance. Further off, spinner dolphins caper in the crystalline waters of the lagoon sheltered by the coral reef. The turquoise shallows lapping the sparkling white shore glow with such unbelievable intensity that it is almost painful to look at.

All the while, albatrosses are streaming in from the high seas, their underwings reflecting pale emerald against the lapis lazuli sky. It is February and hatching is nearing completion among the seething mass of exuberant birds. Everywhere small cream and grey chicks with spiky hairdos sit in bowl-like depressions that pass for nests. They use their little paddle feet to scrape out neatly rounded bowls in the powdery white sand in order to better accommodate their bulging round bellies. I can almost see them growing before my eyes as they constantly beg for — and receive — great gulletfuls of food from doting parents.

For every albatross pair with a chick there are

at least as many non-nesters filling the colony with their animated antics and passionate courtship rituals. Some, finding companionship in their growing pair bond, sit dozing, pressed close together, peach-coloured beaks and ash-dusted cheeks touching tenderly.

The Laysans are among the smallest of all albatross, which also makes them highly agile. From dawn until long after dusk, and from my bedroom window to the far end of the island four kilometres (2.5 miles) away, a quasi-melodious albatross symphony combines whistle-like notes with a bill-clapping tempo resembling castanets. When relative calm settles over the island late in the night, flocks of Christmas shear-waters drift in from the starlit skies, their hushed songs spreading from burrow to burrow.

Black-footed albatross

Around the periphery of Midway Atoll, along

ABOVE Toddling chicks sometimes stray away from their own nest, settling temporarily with a neighbour, where the returning parent doesn't distinguish the visitor from its own.
LEFT A newly hatched chick receives its first meal.

ABOVE Kicking up fine coral sand, a Laysan albatross gathers speed by running down the beach for take-off.
RIGHT The characteristic hunched gait of the black-footed albatross.

the immaculate sand berms and fleshy-leaved native *Scaevola* bushes, nest the island's 'other' albatrosses, the black-footed — though why an almost all dark bird should be named specifically after its black feet remains a mystery to me. Somewhat larger than the Laysans, with only a powder-like dusting of white around the base of their bills, the black-foots cut striking figures in their icing-sugar surroundings. Not so densely packed and overall far less numerous than their smaller kin, they sit with poise and tranquillity, outlined against the swimming-pool blues of the lagoon. Their oddly pale chicks, much lighter than Laysans, seem to match the sunlit environs. Many of the larger black-foot chicks are already on their own while both their parents are quartering the ocean in search of their favourite food, clusters of floating fish eggs. Their feeding range extends all the way to the coast of California.

An adult lands near the tideline and walks purposefully to its nest to deliver just such a treat.

LEFT Left on its own while its parents are away feeding, a black-footed albatross chick scrapes out a comfortable hollow in lieu of a nest in the soft sand.
BELOW A beakful of fish eggs attached to small pieces of flotsam make a typical meal for a black-footed chick, but increasingly indigestible plastic debris is included, revealed in dead chick remains.

The lucky chick struggles to down the dense, rich meal of caviar-equivalent. But there is a hidden tragedy lurking here. Fish egg clusters are often attached to small bits of flotsam, and in today's world much of the North Pacific flotsam consists of plastic and other refuse. As they swallow the eggs the birds also ingest the debris, which they then pass on to their chicks. Unwittingly, both black-foots and Laysans pump their chicks full of this indigestible garbage. As a result, the albatross colony is littered with small colourful mounds of our disgusting cast-offs — cigarette lighters, toothbrushes, light bulbs, bottle caps, plastic soldiers and other kiddies' toys; the list goes on and on. Looking closely reveals each little pile is surrounded by pathetic desiccated scraps of bones and down, all that is left of countless baby albatrosses who never flew. Chillingly, thus far studies have been unable to prove whether the stomach clutter was the cause of death or not, for the simple reason that it is impossible to find

Spirits of the Oceans Wild • 117

ABOVE The courtship between black-footed albatrosses reaches a crescendo as dusk brings freshness to Midway Atoll.
RIGHT A courting pair arch their necks gracefully.

uncontaminated birds to compare respective health levels and survival rates. In the early afternoon when the sun is most intense and the rest of the island seems to be dozing, red-tailed tropicbirds dance an aerial version of the twist above the glowing beaches. Later, when the afternoon freshness returns and delicate patterns of small cumulus clouds light up with sunset colours, activity throughout the albatross colony is energised once more. Not only are more parents returning to care for their chicks, but the many 'loafers' who for whatever reason are not nesting this year — those who've lost eggs, chicks, or perhaps mates, or who are simply still in their adolescent years seeking out a partner — launch into frenzied courtship displays.

Of all albatross dances, never have I witnessed such an utterly soul-lifting routine as that of the black-foots. They arch their necks and cross their beaks; they lift their wings like bouffant shoulders, calling and nodding; they throw their heads high, then

ABOVE Away from the busy colony, a couple find relief from the bright noon glare beneath *Scaevola*, or sea grape, growing at the edge of the warm coral reef lagoon.

bow to touch their shoulders in ritualised preening gestures. Gracefully they twirl and pirouette, sway and swirl in something like an avian version of a flamenco dance. Round and round they go, their webbed feet rising up on tiptoes to the cadence of some inaudible inner music — a truly mesmerising spectacle.

Short-tailed albatross

Among all the hundreds of thousands of feathered inhabitants of Midway, on my third day a single much larger bird suddenly grabs my undivided attention. Though the same basic shape and form as both black-footed and Laysan albatrosses, not only is it substantially bulkier, but its exquisite colours and shades are totally distinct. This is the revered golden gooney, or short-tailed albatross. A single female, she has been visiting this same spot each winter for a number of years and once even laid an egg. Her lonely vigil is a wistful statement for a species literally back from the dead.

The story of the short-tailed albatross is one of the saddest sagas of blood-curdling abuse and mindless destruction of our natural world — yet with a happy ending. Many scientists believe that this species was once the most numerous albatross in the world, nesting by the hundreds of thousands, or possibly even millions, on islands throughout the North Pacific. Fossil remains also indicate that some 400,000 years ago a thriving colony even existed on Bermuda Island in the western Atlantic. But the species' fate took a frightful turn with the rise of the feather-gathering industry that also affected the two Hawaiian albatrosses. In 1889 an eyewitness account of the carnage on the species' stronghold of Torishima Island, an active volcano south-east of Japan, gives us a chilling description of the efficacy of the hunt. The birds on their nests were so steadfast that a man armed just with a club could dispatch up to 200 albatross per day, removing roughly 100 grams (3 oz) of prime feathers from each.

ABOVE Their courtship resembling an avian version of a classic waltz, black-footed albatrosses spin, rise and fall in rhythmic pirouettes.
ABOVE RIGHT The beautiful, rare short-tailed albatross, which nests only in the north-western Pacific, is making a valiant comeback after its very close call with extinction in the mid-1900s.
FAR RIGHT A lone short-tailed female banded on Torishima Island off Japan, has frequented Midway Atoll in Hawaii for many years, trying to attract a mate amongst the local black-footed albatrosses.

At one point some 300 people were employed in the trade on Torishima alone, yielding nearly 40 tonnes of feathers per season, the equivalent of about 300,000 dead birds. Needless to say, at least half that number of chicks also died when one or both their parents perished. This incredible massacre went on for some five decades, and by 1932 the last of the short-tailed colonies lay utterly silent. Few people even knew what the world had lost, and only a handful of ornithologists continued to search for survivors. Eventually they became resigned to the tragedy of their extinction as not a single sighting was reported during the following two decades.

Then, in early 1951, a biological miracle occurred. As if an apparition materialised from another planet, 10 fully adult short-tailed albatrosses were sighted on the black volcanic scoria slopes of the rumbling volcano where the last of their kind had perished all those years ago. Evidently a few early fledglings had managed to depart the island in that doom year of 1931. For a full 20 years they had roamed the ocean unseen. Perhaps they visited the uninhabited island from time to time, but not finding any of their kin, did not stay long.

Finally, as they matured, this tiny gaggle began to lay a few eggs — seven in 1954 — and though many of them failed to hatch in the decade that followed, eventually some chicks were raised and fledged. Thirty years after their rediscovery, the first of those chicks found a mate and also began to breed, so launching a new generation. In the 25 years since then, this most tenuous of recoveries finally began to gather momentum, and today there are about 400 breeding pairs, a few even sprinkled on a couple of their other ancestral islands, such as Senkuku Retto near Taiwan.

For a long time I watch the lone female on Midway, her deep black eyes revealing little of her past. She rests peacefully among the throngs of her busy relatives, and every once in a while tries

unsuccessfully to engage them in courtship. She raises her bill and lifts her wings, moaning loudly, but her baffled entourage seems intimidated rather than enticed, and steps back. A large band on her leg attests to the fact that she once came from Torishima, and no explanation can be found why she doesn't return there. In my mind, I try to imagine her as a survivor of those great massacres of yore. She could still be one of those chicks who came into this world among the last of the nesting multitudes before they were annihilated 75 years ago. Perhaps her earliest chick memories are of the horizon around her nest obscured by the hubbub of the colony, which she now tries to find amid the thriving Laysans and black-foots.

Her beautiful white plumage, the brush of gold and russet on her head and neck, her delicately pink bill offset by a black base trim and a powder blue tip, all imprint themselves profoundly on my mind, aware as I am that we have come so very, very close to never beholding such beauty ever again.

Hawaii's Midway Atoll is a hubbub of sound and movement, where Laysan and black-footed albatrosses intermingle, their fast-paced courtship a mesmerising sight.

8
Under the Tropical Sun
The Galapagos Albatross

Waved albatross

The air is heavy and still, steamy with salt spray and recent torrential downpours. Long, lazy swells, travelling up from the far south, slowly heave the tropical Pacific against unyielding basaltic cliffs. Untrammelled by the modern world not so far away, the mood on Española Island, the southernmost of the Galapagos and only a few dozen kilometres south of the equator, is both peaceful and primordial. Normally rugged and austere, the island seems almost genteel at this time of year. The brief rainy season is already coming to an end, having dispelled the brooding inversion layer and quelled all vestiges of the steady trade winds that blow here much of the time. The arid boulderfields have turned into the nearest thing to a flower-studded meadow this island will ever know.

April is also the arrival time of the waved albatrosses, named after the thin wavy lines on their chest and shoulders. I can see them gathered just offshore in dense rafts undulating like magic carpets, their white heads glowing in the pale late-afternoon sun. Wave after wave, they lift off by the scores, wings labouring as they run down the face of the swells to gain speed. Catching the slight breeze they bank over the cliff line and circle above the low-lying plateau, eventually joining the ranks already on the ground.

As the cool season returns to Española so do the albatrosses. Though moderate-sized by albatross standards, weighing between three and four kilograms (6.6–8.8 lb), they stand proud, reaching nearly a metre (3 ft) at a stretch. Their tall legs, slender necks and proportionately much longer beaks than other species lend them an oversize presence. They are busy reclaiming their traditional nesting areas along the windward side of the island, their only breeding ground apart from a few stray pairs on Isla de la Plata near the coast of Ecuador. They will be here for the next nine months, until the

ABOVE Simultaneous gaping is a frequent feature of the waved albatross courtship.
OPPOSITE Landing on the dark lava shore of Española Island, a waved albatross sways as it walks towards its mate a short distance inland.
Photo: Mark Jones.

ABOVE The waved albatross of Española Island in the Galapagos is the only truly tropical albatross species.
ABOVE RIGHT After spending the warmer hours of the day rafting in large groups offshore, a waved albatross returns to the colony to rejoin the courtship activity.
RIGHT Nesting begins just after the tropical rains have ended, in short-lived meadows where *Tribulus* flowers soon wither to sharp-spined burs.
FAR RIGHT Extremely rapid, loud bill clapping is an unusual and striking courtship feature. Photo: Mark Jones.

onset of the next hot season. The waved albatross is the only truly tropical representative of the entire albatross family.

The colony is a hubbub of activity, and with the sinking sun relieving the oppressive heat of the day, the pace quickens. More and more birds are landing, not the most graceful exercise as they attempt, often unsuccessfully, to slow down from their high-speed glide to a controlled stall and touchdown. Undeterred by ungainly crashes among the boulders, the new arrivals begin investigating potential partners almost immediately. They approach one another with a highly exaggerated swaying gait which appears to say, 'I want to dance', and soon launch into enthusiastic courtship sessions, sometimes in duos, trios or even quartets. With frenetic speed and code-like synchrony, they bow and nod, sky-point, duck, bill rattle and fake preen. As I watch transfixed, I begin to see patterns emerging from seemingly random

and erratic moves. The entire routine is actually a closely choreographed sequence of precisely timed gestures, as if to test their coordination, performed equally by both the male and the female. For example, intense 'fencing' bouts, with billtips clicking and circling around one another, lead to a sudden freeze, standing bolt upright with beaks gaping widely, sometimes simultaneously, sometimes in turn. Each time one bird tucks its beak to its shoulder in a fake preen, the other stretches his or her head down low and claps its beak so fast it makes a sharp rattling sound. On this cue, the first bird swings its head high in a vigorous arc and snaps its bill loudly: 'Clop'. The whole performance is punctuated with screams, moans, sighs and strangely modulated trumpeting sounds — a truly enthralling natural ballet performed against a backdrop of stark beauty.

Some pairs, having completed a whole dance session without missing a beat, finally conclude

Spirits of the Oceans Wild • 127

ABOVE In a rough and tumble attack, a male waved albatross on Española Island rushes a newly arrived female, knocking her to the ground to mate.
RIGHT Delicate mutual preening is a strong part of pair bonding.

with repeated bowing focused at the ground as if inspecting an egg. Exhausted, they settle down together, dozing cheek to cheek, or preening each other's heads and necks ever so delicately. In startling contrast, there is another albatross show going on in parallel to all this ritualised, orderly demonstration of pair bonding, one where chaos and violence seem to reign. For no apparent reason, wild chases and loud fights erupt regularly here and there as males attack perceived rivals. But even stranger happenings take place in the grassy 'landing strip' areas near the cliff edge. Here a number of loitering males are waiting to ambush the females that drop in, whom they promptly rush and pin down in what looks every bit like rape. Many females evade these attacks as best they can, but others appear to partake willingly. The outcome of this promiscuous behaviour will be that as many as one in four chicks will not have been fathered by the male raising it.

LEFT A female retaliates, pushing off the male in a forced mating. Photo: Mark Jones.

ABOVE To gain enough speed for take-off on a windless day, a departing albatross paddles furiously across a boulder field at Punta Cevallos on Española Island. Photo: Mark Jones.

While more birds arrive in the colony many eggs have already been laid, with females sitting resolutely for the first incubation stint while their mates depart to feed at sea. For 62 days the pair will take turns, relieving each other every three weeks or so until the chick hatches. But once again the waved albatross is the odd one out. Unlike all other species, it builds no nest whatsoever, depositing its egg directly onto the uneven rocky ground. Inexplicably, the incubating bird then proceeds to shuffle about with the egg held tightly between its tarsi (lower leg bones), shifting around for no apparent reason, sometimes travelling a considerable distance from where the egg was laid (see Chapter 15). As a result, dozens of eggs become irretrievably wedged in fissures and between boulders.

Many more eggs are abandoned each season when adults give up the task as they succumb to excessive heat or, if the rains run late, dense swarms of mosquitoes. During El Niño events, when exceptionally warm waters waft down from the Panama region, abandonment can reach 100 per cent and for an entire year only bleached eggs may reveal the existence of the waved albatross in Galapagos.

It is clearly not easy being an albatross nesting on the equator. The only reason the waved albatross can do so is by relying on the extraordinary productivity of the Humboldt Current welling up from the deep ocean trench along the continental shelf of Peru, with feeding trips often exceeding 3000 kilometres (1864 miles). Indeed, enormous numbers of other seabirds — boobies, pelicans, cormorants and terns — reside year-round in these fertile waters, nesting in vast colonies on the famed Guano Islands of the region. Why the albatross has not joined them there can, in my mind, only be explained by the long association that pre-Inca coastal civilisations had with these islands, going back thousands of years. Then, as now,

ABOVE Courtship on Espanola Island is most intense in the golden hour just after sunrise and just before sunset.
LEFT An egg begins to hatch after two months of being rolled precariously on the ground, with parents taking great care in swapping incubation duties.

Spirits of the Oceans Wild • 131

ABOVE Heavy El Niño rains sometimes cause the entire waved albatross colony on Española Island to be abandoned if the cooler tradewinds do not return promptly at the onset of the nesting season.
RIGHT Chicks are fed regularly when small.

albatross reproduction strategies could not withstand the depredations of human harvesting.

Perhaps in response to these pressures, waved albatross opt for enormous commutes to bring food back to their chicks ensconced on a sun-baked tropical island where people have never settled. This is a tenuous lifeline. When the chicks are still small and cannot be left alone for long, the parents scour the waters of the Galapagos Marine Reserve instead, but pickings are less abundant here. Although safe while nesting and feeding in these islands, fishing in another country's waters is proving dangerous business, as fishermen from coastal Peruvian villages are finding the big birds quite appetising. Converging on the Humboldt Current's riches from as far away as New Zealand, clear across the opposite side of the Pacific Ocean, the Chatham albatross sometimes shares this sad fate.

Two months after my April visit to Española Island I am back again with scientist friends who are conducting long-term research. This is a good year for the waved albatross.

Plump coffee-cream and chocolate chicks sit everywhere, blending into the dull colours of the once-again arid landscape. Many parents are still guarding the younger chicks, but the bigger ones are already on their own. In the absence of nests, they too shuffle about, dragging their little potbellies along the ground as they seek shade beneath dense thorn bushes. Maybe these private spots also help them evade predation from the resident pair of Galapagos hawks who survey the area from atop a nearby lava hill.

One such wandering chick has taken up temporary residence in the middle of the scientists' rustic open-air camp. Today beady-eyed 'Lincoln' is comfortably installed between stacks of canned beans and peas. His mother, known in camp as 'Snappy Trudy', is home with him while her mate 'Albert' is away at sea. By evening, Mum is gone and Lincoln has moved to a sandy patch underneath the driftwood table, so we must be careful where we place our feet at dinnertime. Living in tune with these extraordinary birds is just one of the many little wonders bestowed upon those who work to save albatrosses around the world.

ABOVE LEFT The low, grey skies of the cool season, when a strong inversion layer hangs over Española Island, indicate a good year for a newly hatched chick guarded by one parent while the other is away gathering food.
ABOVE RIGHT Hatching on the bare ground without a nest, a young chick is fussed over by its attentive parent.
LEFT At five months, unguarded chicks wander off to seek shade in dry vegetation, emerging at the sound of a returning parent's call.

LEFT A mate's perspective of the gaping display is an impressive sight. Photo: Mark Jones.
FAR LEFT Able to soar in light winds, a waved albatross returns to Española Island from the cool upwellings of the Humboldt Current off Peru during a late rainy season shower.
BELOW With the largest bill in proportion to body of any albatross, the waved albatross seems to centre much of its courtship around this striking feature, with fencing, rattling, clapping and gaping performed in rigorous chronology.

Spirits of the Oceans Wild • 135

Part Two

Science and Conservation

Mark Jones

*I am the albatross that awaits you at the end
    of the earth.
I am the forgotten soul of the drowned sailors
    who rounded Cape Horn from all the oceans
    of the world.
But they did not perish in those furious waves.
Today they fly on my wings, towards eternity,
    within the ultimate sanctum of the Antarctic winds.*

— Sara Vial, Chilean poet, 1992.
(Words on the albatross monument of Cape Horn;
free translation from original Spanish by Tui De Roy.)

9
Perspectives:
Albatrosses and Man through the Ages
Mark Jones

*Mad as the sea and wind, when both contend
Which is the mightier...*
— William Shakespeare, *Hamlet*, 1601.

Shakespeare, for all his fabled storms, raging tempests and love of lyrical metaphor, and unlike his literary successors, makes no mention of the albatross in any of his prolific musings. In William's time, the bird, much less the word, had not become part of the vernacular.

But the world back then was beginning to shrink. Voyage after exploratory voyage during the fifteenth, sixteenth and seventeenth centuries — the so-called Age of Discovery — systematically eroded geographical and natural unknowns, and redefined the parameters of the legendary Seven Seas. In the century before The Bard's first plays swept the Elizabethan stage, a succession of Portuguese mariners had forged their way about the Atlantic, mapping the winds of the south and gradually charting the African coastline until, in 1488, Bartolomeu Dias succeeded in rounding the continent's southern cape, so discovering the vital sea route to the riches of the Indian Ocean. Thus they became the first Europeans to venture into the blustery domain of the albatross. The word Alcatraz, a Portuguese adaptation of an Arabic term for pelican liberally applied to a number of large seabirds, is the root for what became the evolving, quasi-universal word for those greatest of all marine birds: *Al-câdous... Alcatráz... Alcatrasses... Algatrosses... Albitrosses... Albatross...*

Tristão da Cunha, in discovering the lone volcanic island that now bears his name while sailing down the middle of the South Atlantic in 1506 (see Chapter 20), would have been among the first sailors to see an albatross breeding colony — though bird-watching per se was surely not within his royal charter! Fourteen years later, fuelled by the flourishing capitalist culture of a prospering

ABOVE As voyages of exploration pushed south, a wandering albatross gliding effortlessly past taut sails epitomises the beginnings of an uneasy relationship. This photo was taken by Frank Hurley, who joined both Mawson's and Shackleton's famous Antarctic expeditions in the early twentieth century.
OPPOSITE Royal albatrosses on Campbell Island gather in a lively 'gamming' session, a term referring to seafarers' reunions.
PREVIOUS PAGES Buller's albatrosses fill the evening sky over The Forty-Fours off the Chatham Islands.

60. The wandering ALBATROSS in the INDIAN OCEAN. July 1842.

ABOVE Painted from the deck of HMS *Fly* in 1842, wandering albatrosses surround a tall ship sailing through the Indian Ocean on its way to Australia.
OPPOSITE TOP From the pages of a diary kept in 1842 during a voyage to New Zealand aboard the *Clifford*, a royal albatross caught on a fishing line struggles for its life.
OPPOSITE BOTTOM Wanderers attracted to discarded food have been at risk since the first sailing vessels began plying the southern oceans centuries ago.

Europe, and the fierce rivalry that ensued between nations intent on monopolist empires, yet another Portuguese navigator, though this time bearing the Spanish standard, sailed into the Pacific Ocean for the first time. Ferdinand Magellan and the wretched crews of his four caravels were undoubtedly escorted by hordes of foraging black-browed albatross as they transited the straits that were to become the gateway to the lucrative Pacific trading routes.

And so, with one of Magellan's ships going on to complete the first circumnavigation, the curtain was raised on a stage set for future navigators, explorers and exploiters to ply the southern seas — the patent realm of the albatross.

> *I protest before God and as my soul shall answer for it that I think there were never in any place in the world worthier ships than these.*
> — Charles Howard, Lord High Admiral of England, 1588.

In the foaming wakes of the Portuguese and the Spaniards, who by papal bull (Pope's decree) had agreed the world beyond Europe should be divided into two exclusive shares, sailed the British, the French and the Dutch to contend for maritime supremacy. Hence, the ships of zealous privateers — paid pirates on government-sanctioned missions — along with well-armed royal navies, and finally profit-driven trading companies, all began to regularly ply the southern routes. More and more seamen came into contact with the strange animals and birds that inhabited those wild, tempestuous regions. In their official logs and fo'c'sle yarns, amid the elaborated tales of bedevilled sea monsters and mythical sirens, they brought home wondrous but truthful stories of great 'seaffowle', 'big as swannes' riding the incessant gales and defiantly skimming the wavetops.

A typical voyage — if one survived the privations — was three to four years. I find it not surprising, then, that a majestic bird that would fly alongside a ship, sometimes for days on end, habitually reappearing through fog and hail, come storm or shine and as far from land as a sailor can go, would etch itself into the psyche of even the toughest seaman. Those crowded decks were brutal and rowdy confines, and the men, rough, unruly and chiefly illiterate, thrived on superstition and lore.

140 • Science and Conservation

It's difficult nowadays to fully imagine what it would have been like to sail the oceans at that time. The limitless horizons foretold of discoveries yet to inscribe the charts. The seas were pristine, and the abundance and diversity of nature intact. On a calm night aboard a creaking barque, during the forever-darkness of the middle watch, and with just the background thrum of the rigging straining at the fastenings in cadence with the rhythmical scend of a tranquil ocean, the sudden blows of a pod of whales surfacing alongside would split the night air with the audacity of gunshot. The raucous cries of an unseen flotilla of penguins, or the melancholy mewings of great flocks of dabbling petrels would have been an almost constant and familiar symphony in the far reaches of the southern seas. On other days, furious storms would have instilled sheer terror into the souls of men, cursing as they hauled the braces and scaled the flailing masts of a wildly careening galleon, clutching at sodden, flogging canvas, their very lives at the mercy of the shrieking winds and unrelenting seas... And there, with the grace and composure of an angelic apparition, a stately albatross would soar through the maelstrom, unperturbed and reigning... Little wonder that the foundered souls of their comrades were reincarnated in the serenity of those arcing wings: man's intimate relationship with the bird of legends had taken flight in no uncertain terms.

> ... dark clowdy weather ... fresh gales ... with some rain, PM had a boat out and shott several sorts of Birds, one of which was an Albetross as large as a Goose whose wings when extended measured 10 feet 2 Inches, this was grey but their are of them all white except the very tip end of their wing...
> — Lieutenant James Cook, Journal, 4 February 1769, one week after rounding Cape Horn.

Discovery followed discovery and the oceanic provinces of the albatross gradually became human conquest. Amsterdam Island, which today harbours the most endangered species of them all, was the first subantarctic island to be sighted in 1522. In 1578 the Drake Passage was opened up as the way around Cape Horn, and the storm-battered Diego Ramirez Islands some 30 kilometres (18 miles) south-west of the island-cape — and home to the southernmost breeding albatrosses, with colonies of grey-headed and black-browed — was discovered in 1619. One by one, from the far reaches of the Indian Ocean to beyond the South Pacific, forlorn dots of land made their way onto rudimentary charts: Marion Island sighted by the Dutch in 1663, South Georgia in 1675, Prince Edward, Crozet and Kerguelen in 1772 by the French. Albatrosses accompanied the ships, and along the way hungry sailors eager for a square meal '...made a provision of their meat...'. Ships' logs became liberally strewn with records of birds caught to satiate the men.

James Cook, spectacularly accomplished and justly ambitious, hailed as 'The ablest and most renowned Navigator any country hath produced', is credited with changing the face of the globe more than any other seaman. From Newfoundland to New Zealand, Tahiti to the Aleutians, his meticulous maps and sketches formed the backbone for the evolution of the modern charts. During three successive circumnavigations, and by sailing beyond the Antarctic Circle, he finally quelled the centuries-old belief in *Terra Australis Incognita*, an unknown expansive southern land mass first speculated by Aristotle to 'balance' those of the northern hemisphere. In fulfilling his self-ascribed charter to go not only '...farther than any man has been before me, but as far as I think it possible for man to go...' he pinpointed islands only vaguely known to exist. The time he spent in high seas albatross territory is even today perhaps only rivalled by modern Southern Ocean fisheries skippers.

Excerpts from *The Rime Of The Ancyent Marinere*
— Samuel Taylor Coleridge, 1798.

It is an ancyent Marinere,
And he stoppeth one of three:
"By thy long grey beard and thy glittering eye
"Now wherefore stoppest me?

Listen, Stranger! Storm and Wind,
A Wind and Tempest strong!
For days and weeks it play'd us freaks—
Like Chaff we drove along.

At length did cross an Albatross,
Thorough the Fog it came;
And an it were a Christian Soul,
We hail'd it in God's name.

And a good south wind sprung up behind,
The Albatross did follow;
And every day for food or play
Came to the Marinere's hollo!

"God save thee, ancyent Marinere!
"From the fiends that plague thee thus—
"Why look'st thou so?"—with my cross bow
I shot the Albatross.

And I had done an hellish thing
And it would work 'em woe;
For all averr'd, I had kill'd the Bird
That made the Breeze to blow.

Day after day, day after day,
We stuck, ne breath ne motion,
As idle as a painted Ship
Upon a painted Ocean.

Water, water every where
And all the boards did shrink;
Water, water every where,
Ne any drop to drink.

Ah wel-a-day! what evil looks
Had I from old and young;
Instead of the Cross the Albatross
About my neck was hung.

How long in that same fit I lay,
I have not to declare;
But ere my living life return'd,
I heard and in my soul discern'd
Two voices in the air.

"Is it he? quoth one, "Is this the man?
"By him who died on cross,
"With his cruel blow he lay'd full low
"The harmless Albatross"

The other with a softer voice
As soft as honey-dew:
Quoth he the man hath penance done,
And penance more will do.

Farewell, farewell! but this I tell
To thee, thou wedding-guest!
He prayeth well who loveth well
Both man and bird and beast.

Bird Island, one of the most intensely studied albatross colonies anywhere (see Croxall, Chapter 13), was named on Cook's 1775 hand-drawn chart of South Georgia Island. The astronomer William Wales, on that second voyage when *Resolution* repeatedly pressed south towards the pack ice, happened to become the tutor of a budding young philosopher and poet destined for literary prominence. It is speculated that the tales that Samuel Taylor Coleridge must have heard recounted by his worldly-wise teacher may have sown the ideas behind his supernatural, enigmatic epic *Rime of the Ancient Mariner*. Thus, in 1798, the word albatross became firmly planted into the language and consciousness of the pre-Victorian era. The notions Coleridge — who never saw an albatross — instilled about the exalted status of the bird have since ricocheted through the generations and entered the everyday lexicon.

> *Yet whales and seals, petrels and albatross, are exceedingly abundant throughout this part of the ocean [between 56 and 57 degrees south]. It has always been a mystery to me on what the albatross, which lives far from the shore, can subsist...*
> — Charles Darwin,
> The Voyage of the Beagle, 1839.

Long before our forebears speculated on the shape of the earth, the balance of its land masses and the nature of the universe, albatrosses were circumnavigating the globe, riding the perpetual winds and waves, and navigating flawlessly to the isolated specks of land where they could breed. The earliest fossils trace their nascent record some 50 million years, with recognisable petrels and albatrosses already plying the prehistoric seas some 32 million years ago. They evolved in a steadily changing world, adapting to modified climate patterns and the shifting dynamics of primordial oceans. As continents drifted apart to occupy their present positions, ice ages came and went, sea levels fluctuated and currents wavered, and the productivity of the seas ebbed and flowed in response. Fossils from southern land masses confirm that genera very similar to today's albatrosses lived in the Southern Ocean as early as nine million years ago, and throughout the North Pacific even before hominids began strutting about the plains of Africa. Although Pleistocene (1.8 million to 10,000 years ago) fossil breeding sites of short-tailed albatross have been unearthed in Bermuda, several other species had already disappeared from the North Atlantic after the Isthmus of Panama closed off the Caribbean-Pacific connection and shut down the currents that had spanned the equatorial seas. Gradually the distribution of albatrosses and the highly productive waters in which they foraged settled into that which we know today. These constantly evolving systems will continue to shift in eons to come. But nowadays there is an added factor in the equation: we humans have but recently brought new pressures to bear on these fine, naturally balanced patterns, casting serious and threatening question marks over the viability of their life-sustaining cycles. And the crux is that this is now happening at a vertiginously accelerating pace.

Only a few of our ancestral peoples would have been familiar with albatrosses. Southern African and Australian aboriginals almost certainly encountered albatrosses around their shores. In the North Pacific, midden remains left by the ancients who settled the Aleutian Islands and migrated down the west coast of North America attest to their dietary preference for the large short-tailed albatross. By the time the human migration ended at the tip of the South American continent, the tall Patagonian Indians whose constantly burning fires gave rise to the name of 'Tierra Del Fuego' (Land of Fire), were very accustomed to albatrosses careening through the convoluted waterways between the myriad islands they inhabited. Likewise Asian fishers venturing off Japan and the East China Sea would have regularly encountered the albatrosses that breed in the nearby western Pacific islands. And, finally, Polynesian voyagers heading towards Hawaii, and especially those that bore southwards to discover the distant islands that were to become New Zealand, penetrated into the very heart of albatross country. After the Aleuts, these native seafarers probably made the most use of albatrosses. Their Moriori and Maori descendants eventually settled the Chatham Islands, making regular seasonal raids to the outlying albatross colonies for both staple and ceremonial foods, a tradition that lingered through to modern times. Albatross bones and feathers have been utilised for everything from chiefly adornments to fishing lures. Even so, relative calm reigned throughout the albatross world until club and spear were usurped by the crack of gunshot heralding the arrival of the first Europeans.

> *Calm again: went out and shoot Diomedaea Exulans Albatross or Alcatrace, differing from those seen to the Northward of Streights of La Maire in being much larger and often quite white on the back between the wings, tho certainly the same species.*
> — Joseph Banks, Journal, 500 nautical miles south-west of Cape Horn, 3 February 1769.

It had become fashionable from Magellan's day onwards for captains to carry on board scholarly men and artists to keep meticulous journals and make detailed observations of the geographical and natural phenomena they came upon, and to amass preserved specimens for eager collectors and museums. Among these voyaging scientists and naturalists are many of our most famous historical figures: Banks, Hooker, Forster, Rothschild, Darwin, Gould. Individual species were described in detail and the hierarchy of nomenclature defined, though some of the confusion they created with descriptions of such similar-looking species has reigned through to modern times. Today, albatross taxonomy is still hotly debated and evolving

ABOVE Compared to a human hand, web prints of a southern royal albatross in a Campbell Island peat bog give an idea of the bird's gigantic proportions.
OPPOSITE Details of shipboard life aboard the *Duke of Portland*, sketched during the long passage from England to the new colonies, reflect the constant presence of albatrosses in southern latitudes, including at least one unfortunate bird brought onto the vessel's decks, samples of its feathers preserved amongst other voyage memories.
FAR LEFT In Coleridge's epic poem, as sole survivor of a doomed ship, the ancient mariner faces the eternal curse of a dying crew for shooting an albatross, living his penance by teaching others about caring for all creatures of our planet. These stanzas excerpted from the 149-verse original version were first published in *Lyrical Ballads* in 1798. The full e-text, courtesy of Risa S. Bear of Renascence Editions, can be viewed at www.uoregon.edu/~rbear/ballads.html.

with forensic precision (see Part Three for individual species accounts).

All albatrosses belong to the order of the Procellariiformes, which incorporates all the 'tube-nosed' seabirds — their single most distinctive feature — encompassing both the petrels and albatrosses. Divided into four family groups, they represent the entire gamut of largest and smallest of all seabirds, as well as the most numerous, both in terms of species and abundance, or biomass. Thus we have the family Diomedeidae (albatrosses), the Procellariidae (petrels and shearwaters), the Hydrobatidae (storm petrels) and the Pelecanoididae (diving petrels), some 140 species all told. Distributed throughout the world's seas and ranging to the highest latitudes, the Latin root of their name, Procella, proclaims their love of violent storms.

> He [Kotick, the white seal] picked up with an old stumpy-tailed albatross, who told him that Kerguelen Island was the very place for peace and quiet...
> — Rudyard Kipling, *The Jungle Book*, 1894.

By 1854 all of the subantarctic islands had been discovered. Detailed accounts of those distant isles thronging with fur seals, their surrounding seas so teeming with whales there was supposedly scarcely room to manoeuvre a ship, had grabbed the attention of profiteering sealing and whaling consortia on both sides of the Atlantic. The Industrial Revolution was by then well under way and the burgeoning European and American economies were needy of whale oil and furs, not to mention new lands to colonise. Fortunes were to be made, and battles — both on the water and in the boardroom — were being fought over them. Consequently, many ill-equipped seamen were shipwrecked or marooned, and where they could they eked a meagre living off the wildlife they encountered. Or they perished. Albatross eggs, chicks and adults alike were harvested like fowl, and there was many a harrowing tale of survival. The term gooney, or goney, arose for birds acting trustingly enough — stupid to the sailors' way of thinking — to allow themselves to be captured. During the sealing era on the Bounty Islands a makeshift hovel was even thatched with albatross wings.

Also, as more and more vessels carried burgeoning numbers of settlers destined for new lives in colonial lands, shooting albatrosses from the decks became a sport encouraged by captains as passenger entertainment. With hundreds of voyages per year carrying thousands of emigrants to California, Australia and New Zealand during the 1800s, this toll cannot be underestimated.

> Aramis was silent; and his vague glances, luminous as that of an albatross, hovered for a long time over the sea, interrogating space, seeking to pierce the very horizon...
> — Alexander Dumas, from *The Three Musketeers: Man in the Iron Mask*, 1850.

Protracted voyages also allowed for long hours of quiet contemplation of the horizon, with ample time to become captivated by the mesmeric meanderings of constantly reeling birds. Many a fledgling dream has soared across the ocean upon the wings of an albatross. But while some were inspired to wax lyrical in poems and prose, with the albatross etched as a symbol of freedom and mystery, others became more intrigued with their behaviour. At a meeting of the New Zealand Geographic Society in 1868 a Captain Hutton, using his acute knowledge of sailing and mathematics, set out with complex algebraic equations to resolve the mechanical principles of the 'problem of albatross flight', which he surmised was 'opposed to the known laws of physical science'. The albatross's perfect mastery of the complex kinetics in the art of dynamic soaring is indeed one of nature's wonders to behold. I find it ironically reassuring that in today's crisis-managed world of traded carbon deficits and unsustainable energy demands, for a gulletful of squid an albatross may harness the wind for its boundless travels, perhaps the most energy efficient being on earth (see Shaffer Chapter 11).

Despite a history of hunting and the plundering of colonies for eggs and feathers, the southern albatross species have generally fared better than their northern brethren. By far the largest depredations were sustained around the turn of the twentieth century by

BELOW Two Antipodean albatrosses greet as they settle on a windless sea while feeding in the rich upwellings near New Zealand's Kaikoura Peninsula. New Zealand is the capital of the albatross world, with half of all species nesting on its offshore islands, nine of them found nowhere else.

the three North Pacific species on the Japanese and Hawaiian islands. An insatiable demand for plumes on the international market fuelled a lucrative business that, during the boom years, employed up to 300 men and women working a single colony at a time. Over five million short-tailed albatrosses are estimated to have been killed. The term featherweight took on a different meaning, as 383 tonnes of 'swan's down' per annum were exported in the first years of the 1900s. Over a period of 33 years the short-tailed colonies were decimated. So too were the feather hunters themselves when, in a twist of poetic justice, a volcanic eruption on Torishima Island in 1902 — outside of albatross season — wiped out the entire human population of 129 islanders.

Fowling, as it was known, was banned in 1906, yet continued illegally for many years afterwards. The dearth of short-tailed albatross sightings in the middle years of the twentieth century led to the belief that the bird had become extinct. Meanwhile, on the Leeward Islands, US opposition to Japanese incursions into its naval territory curbed some of the exploitation of the Hawaiian Laysan and black-footed colonies, setting in motion the first legislative measures towards the preservation of Pacific albatrosses. Midway Atoll had its own problems with the establishment of the US airbase during the 1930s. Bulldozed and concreted, the sandy atoll became a crucial halfway staging post for transpacific air routes, and bird collisions with planes were a major hazard for albatrosses and aviators alike. For the thousands of military personnel stationed there during the Second World War the babble of courting albatrosses was drowned by the incongruous wailing of air raid sirens and the roar of accelerating aircraft.

While today peace has returned to the colonies, and limited night landings have mitigated the bird-strike issue, the destructive legacies of human invasion linger. Battles of a different kind are now being fought on those isolated shores. The US Fish and Wildlife Service and armies of volunteer conservationists are working to combat swathes of introduced plants that choke prime nesting habitat. Even toxic, lead-based paints peeling from ageing buildings are proving deadly to chicks whose misfortune it is to have hatched nearby.

> ....what have I done to you, I who take as my own everything to the point of destroying all of which my life is not only a part, but also depends?
> — Laure Delvolvé, *Quand les carnassiers n'y sont pas*, 1960.

During the latter half of the twentieth century and into the twenty-first, the incessant drive to feed the world's ever-expanding demand for seafood has by far made the greatest, yet most surreptitious, impacts on albatross survival. The UN Food and Agriculture Organisation (FAO) statistics show

ABOVE Recently uplisted to Critically Endangered status, a rare Tristan albatross soars through a fleeting South Atlantic rainbow.

the 2004 world fishing fleet comprised about four million vessels with 41 million people in the industry. Producing some 140.5 million tonnes, this equates to a mean annual fish consumption of 16.6 kilograms (36.6 lb) *for every human being on the planet*. Nearly 61 per cent of these catch figures — almost 86 million tonnes — represent marine wild fish. Official records indicate that over a quarter of the world's marine fisheries have collapsed through overexploitation, with a further half at or near their limits of sustainability. Development of Regional Fisheries Management Organisations (RFMOs) and the establishment of Marine Protected Areas and reserves are helping to curb the demise of remaining stocks. Clearly, with global profits in the tens of billions of US dollars, fisheries are big business, with China, Peru and the United States leading the pack of industrial producers. This degree of fishing effort leaves no part of the world's seas untouched and extends the brunt of our influence beyond every horizon to the core albatross domain.

It is inevitable that conflicts of interest arise between oceanic wildlife and fishers, each going about their business. Unintentional and unwanted bycatch, the high levels of non-targeted species killed in fishing operations, is one of the sad results, with albatrosses and their petrel brethren suffering losses in the tens of thousands. Actual figures for

Science and Conservation • 145

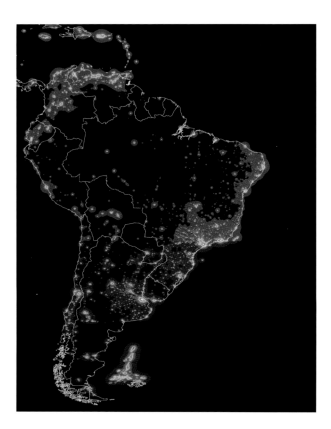

RIGHT Seen from space, massive squid jigging operations on the Patagonian Shelf, which rely on intense artificial lighting to draw their catch towards the surface at night, account for the brightest lights in the southern hemisphere, with unknown effects on albatrosses and other seabirds. Map courtesy of P. Cinzano, F. Falchi (University of Padova), C. D. Elvidge (NOAA National Geophysical Data Center, Boulder). Copyright Royal Astronomical Society. Reproduced from the Monthly Notices of the RAS by permission of Blackwell Science.

this toll vary widely, with annual kill estimates hotly debated in both science and fishers' forums. But the fact remains that steady declines in numerous albatross nesting colonies bear solemn witness to the severe issues of high-seas mortality.

> You can know the name of a bird in all the languages of the world, but when you're finished, you'll know absolutely nothing whatever about the bird.... So let's look at the bird and see what it's doing — that's what counts.
> — Richard Feynman, *What Is Science?* 1966.

Essentially, it's a debate about food: fishing vessels ply the oceans to catch fish and squid, and in so doing they bring to the surface the exact same commodity that, paradoxically, is for a large seabird relatively difficult to capture. The prime reason that albatrosses and petrels range so widely is precisely because for them to catch their prey is not easy. Their access to the ocean's bounty is limited to the depth they can reach from the surface, and, like a gazelle escaping a lion, it is the business of their prey to be fleet, stay away from the danger zone and remain within the safety of the masses to not be caught. Hence, even for a majestic albatross, scavenging is a lucrative, honest way of making a living: everything dies, and many dead things float, so covering vast distances in an energy-efficient manner eventually yields more rewards.

When men took to the sea to go whaling the birds began reaping the benefits of following ships, scavenging a reliable living from discarded offal without paying any dues. A baited longline hook is a different matter: it will float astern a vessel until the weight of the rig drags it down, and during that period it is a lethal, irresistible morsel indistinguishable from any other tasty titbit.

High-seas driftnets were globally banned in 1991 in recognition of their propensity for indiscriminate massacre of marine life, as there was no way to mitigate their effect on untargeted species. This is not the case with longlining, or with trawl fisheries in which birds strike cables or get trapped by haul wires and nets. As I heard one passionate trawl fisher declare, '...this is not a problem about the birds ... it's a problem about our fishing techniques...'. Huge advances have been made in employing various methods to reduce the incidental killing of birds, though the challenge still lies in convincing all fishing fleets to apply these techniques universally. With incentives of lucrative prizes and international awards like the WWF-backed Smart Gear Competition, a number of ingenious and effective devices are being developed specific to individual fishery requirements. The main criterion is to avoid attracting the birds into the vessels' wakes with tempting baits and discarded offal. They can also be generally discouraged from lingering in these danger zones with the vigilant use of tori lines, a bird-scaring device consisting of an array of dangling lines or wands (see Southern Seabird Solutions Chapter 21 and Sullivan Chapter 22). The simple fact remains that virtually every bird in the modern ocean has grown up learning that those noisy hulks on the horizon provide the best chance of a meal. There is also evidence that some populations rely in large part on fisheries discards.

Modifying setting and hauling techniques, and developing bird-friendly catch management are topics tabled at government levels. A bevy of organisations and conventions, international treaties, quotas, observer programmes, education workshops, at-sea task forces and the unbiased sharing of mitigation technology and know-how attempt to span the oceans as widely as the roaming birds. Yet renegade and illegal pirate fishers remain a blight on efforts to legislate the high seas with effective conservation agreements (see De Roy Chapter 26).

Other major concerns fall within regional fisheries like those of South America where the greatest numbers of both fishers and albatrosses converge on the same areas (see Favero Chapter 23). Fears for the future of waved albatross populations feeding in Peruvian waters were recently exacerbated by the revelation of fishermen deliberately catching them as food or bait, and has led to the uplisting of the species to Critically Endangered. With estimates of 59,000 artisanal fishers (near-shore subsistence and/or small-scale commercial fishers) working some 9000 vessels along the Peru coast (2002 figures), this is by no means a minor concern (see Anderson Chapter 15).

...when the last individual of a race of living things breathes no more, another heaven and another earth must pass before such a one can be again.
— William Beebe, *The Bird: Its Form and Function*, 1906.

With time, human attitudes change, perspectives evolve. While once a fashionable quilt was overstuffed with albatross feathers, today it is more likely to depict an Escher-like motif of intertwined albatrosses gliding over a loud acrylic sea. The word albatross has made it into every corner of society: from aircraft types to a computer program, a US navy ship to a golf cart, a publishing house to a popular music group, and hotels from Croatia to New Jersey, Newfoundland to Australia — all these and more proudly sport the name 'Albatross'. And surrounding the bird there is an aura of majestic surrealism that holds more and more people spellbound. Spawned from nature tourism successes in whale and dolphin watching, there is a budding market specifically aimed at intimate albatross encounters (see 'Where to See Albatrosses').

Natural harmony once again reigns at most albatross colonies nowadays. All around the watery globe wildlife-rich islands, once mercilessly plundered, are now among the most rigorously protected and intensely managed lands, kept off-bounds from all but a select few and benefiting the complete diversity of species sharing these unique habitats. Falling under both national regulations and international accords, from Specially Protected Areas to National Parks and Wildlife Reserves, stringent conservation measures, restoration schemes, quarantine processes and intense and massively expensive pest eradication projects are under way. Six island groups where albatrosses breed are World Heritage Sites (the Galapagos, Gough and Inaccessible, the New Zealand Subantarctic, Macquarie, Heard and McDonald, and Mewstone and Pedra Branca as part of the Tasmanian Wilderness Area). In 1957 Japan declared the short-tailed albatross a Special Natural Monument, while under US Endangered Species legislation a fishery bycatch toll of just two of these birds within five years is set to trigger the closure of Alaska's entire US$500 million trawl industry.

Sadly, legacies of alien species such as the introduced house mouse on Gough Island (see Cooper Chapter 24) may still undermine some conservation measures, but gradual successes, such as the eradication of rats from New Zealand's subantarctic Campbell Island, bring substantial gains in safeguarding breeding sites worldwide.

All the world's a stage ...
and one man in his time plays many parts.
— William Shakespeare, *As You Like It*, 1599.

There are no rehearsals for maintaining the integrity of wildlife that shares this planet with us. Urgently addressing such all-encompassing issues as our role in climate change and ecosystem-wide fisheries management is incumbent on today's societies and their leaders. The inactions we espouse today will be the haunting curse we bequeath to our future generations. In many respects albatrosses represent but small collateral damage in the broader schemes of world biodiversity losses, yet as ocean-wide travellers they are true global citizens. Each one of us can play a major role in safeguarding their future. To adapt the phrase from Robert Cushman Murphy, perhaps more of us could say, 'We belong to a high cult of mortals for we have *helped save* the albatross.' If we lose just one species, or simply act as bystander to the demise of a single population, it will be an epitaph hung around the neck of all humanity to wear for eternity, with little prospect of finding a proverbial wedding guest on which to unload the burden.

It is therefore no use to discuss the use of knowledge. Man wants to know, and when he ceases to do so he is no longer man.
— Fridtjof Nansen, *Farthest North*, 1897.

In the following pages of this part of the book some of the world's most dedicated and respected scientists, conscientious fishers and conservation-minded citizens write of their concern for albatrosses or expose the findings of their research, revealing the most outstanding facts and facets of the legendary birds 'That made the Breeze to blow'.

ABOVE When one parent is killed at sea not only does the chick on the nest die, but sometimes the surviving mate remains in attendance, waiting to be spelled by its partner, until it too perishes, as happened to this Buller's albatross on The Forty-Fours.
BELOW Introduced plants on some islands severely impede access to ancient nesting grounds. These courting Laysan albatross will have difficulty taking off from the dense mustard growing on Eastern Island, Midway Atoll.

10
Flagship Species at Half-mast
Rosemary Gales

Dr Rosemary Gales leads the Wildlife and Marine Conservation Section of the Tasmanian Department of Primary Industries and Water. She has worked with marine birds and mammals in the Antarctic and subantarctic, New Zealand, Australia and Newfoundland, and currently is the Chair of the Status and Trends Working Group of the Agreement on the Conservation of Albatrosses and Petrels. Rosemary.Gales@dpiw.tas.gov.au

BELOW AND OPPOSITE
A who's who of all albatross species and their conservation status.

Synopsis: A hard look at the survival status of each species of albatross paints a dark future, but also ignites passion that calls for action while there is still time to avert mass extinction among the most endangered bird family in the world.

'The prospects of survival of the members of the avian family Diomedeidae are more precarious than for most other bird taxa, albatrosses being among the most threatened species of birds in the world.' This statement has been made repeatedly during the last decade as a result of a series of reviews and conservation assessments that have highlighted the widespread population declines and the critical and urgent need for conservation efforts (Gales 1998, Croxall and Gales 1998, BirdLife International 2000).

The reasons for these population declines are numerous and compounding, and largely anthropogenic. Humans have been killing albatrosses since we ventured into their oceanic realm. Following the widespread exploitation by egg and feather collectors, as well as guano traders during the 1800s, our impacts evolved to be less intentional, but equally devastating. In 1993 I first reviewed the global status of albatross populations and the factors affecting them, and concluded that mortality associated with commercial fishing operations had become the most serious threat facing these birds. At that time, the fishing gear most commonly reported to be associated with albatross bycatch was longlines targeting tuna, toothfish and other species. More recently, trawl fishing has also been identified as a threat that rivals that of longline fishing, fisheries which have both flourished since the 1950s. It is the propensity of these birds to be attracted to such vessels and scavenge on baited hooks and discarded fish remains that places them in harm's way. The birds are no less safe when they are ashore, with many populations being impacted by alien species and pathogens.

Meaningful comparisons of the changes in the conservation status of albatrosses over time are sometimes confounded by ongoing taxonomic revisions, the application of new information of population status and trends, and changes in the criteria being applied in the assessment of threatened taxa. In recent years, BirdLife International (BLI) has been responsible for the application of World Conservation Union (IUCN) criteria to identify threatened bird taxa, norms that are widely accepted as objective means of evaluating species at risk of extinction. The criteria are based upon population size and rate of change, as well as extent of breeding distribution and fragmentation.

Wandering Albatross
Vulnerable

Antipodean Albatross
Vulnerable

Tristan Albatross
Critically Endangered

Amsterdam Albatross
Critically Endangered

Northern Royal Albatross
Endangered

Southern Royal Albatross
Vulnerable

Black-browed Albatross
Endangered

Following a taxonomic review of albatrosses in 2006, specialists commissioned by the Agreement of Albatrosses and Petrels (see www.acap.aq), determined that a total of 22 species of albatross should be recognised. Of these, 18 (82 per cent) have a significant threatened status, a stark contrast to the overall rate of 12 per cent for the 9799 worldwide bird species. Albatrosses not only have retained their status as having the highest proportion of threatened species in any bird family that has more than a single species, but have even increased their lead in this unenviable statistic since it was first documented by Croxall and Gales in 1998. Currently, four albatross species (18 per cent) are *Critically Endangered*, six species (27 per cent) are *Endangered*, eight species (37 per cent) are *Vulnerable*, and four species are currently recognised as *Near Threatened*. No albatross species currently warrants the lower category listing of *Least Concern*.

By definition, any species identified as *Critically Endangered* faces an 'extremely high risk of extinction in the wild'. For the **Chatham albatrosses**, this level of risk results from their extremely restricted breeding distribution, an area of less than 10 hectares (22 acres) on The Pyramid, a rock stack off New Zealand's Chatham Islands. Compounding this is a reduction in the condition of their limited habitat as a result of extreme weather and a changing climate. The **Amsterdam albatross** is clearly *Critically Endangered* due to a population of fewer than 100 mature individuals, an extremely restricted breeding area, and the recent likely impacts of disease resulting in elevated chick mortality. In 2007 the **waved albatross** was uplisted to *Critically Endangered* following confirmation of population declines related to mortality in Peruvian artisanal fisheries (Anderson et al. 2002, Awkerman et al. 2006). The **Tristan albatross** joined this category in 2008 due to mouse predation of large chicks on Gough Island (Cuthbert and Hilton 2004), the species' breeding stronghold.

For the six albatross species that qualify as *Endangered*, and hence defined as facing a 'very high risk of extinction in the wild', the current overall population trends are all documented as decreasing. The **northern royal albatross** has an extremely restricted breeding range, with breeding success of the Chatham Island population (99 per cent of all breeding pairs) plummeting due to lack of nest material as a result of storm-induced habitat changes that devastated the vegetation. **Black-footed albatrosses** are at risk because of widespread population declines resulting from interaction with driftnet fisheries in the North Pacific until 1992, and more recently with US and Asian longline fisheries. **Black-browed albatrosses**, the most numerous of all albatross species, are killed in large numbers in trawl fishing operations as well as both pelagic (open sea) and demersal (bottom) longline fisheries throughout their breeding and migration ranges. Population declines of the **Atlantic yellow-nosed albatross** are largely influenced by widespread deaths associated with fishing activities, with some indications that the species may be upgraded to *Critically Endangered* if these threats do not abate. Widespread fisheries mortalities also impact on **Indian yellow-nosed albatrosses**, with population declines at the main Amsterdam Island breeding site exacerbated by the presence of bacterial disease. These diseases may also impact on the **sooty albatrosses** breeding on Amsterdam Island, although there is no information on the trend of this population. The larger populations at Gough Island and the Tristan da Cunha group are decreasing, largely as a result of fisheries, and this species may require listing as *Critically Endangered* if these trends are found to be more widespread.

For the eight albatross species listed as *Vulnerable*, it is their restricted breeding range that is the criterion that most frequently qualifies the species for listing. Reflecting this is the high degree of endemism of these birds, with five

Campbell Albatross
Vulnerable

White-capped Albatross
Near Threatened

Shy Albatross
Near Threatened

Salvin's Albatross
Vulnerable

Chatham Albatross
Critically Endangered

Grey-headed Albatross
Vulnerable

Buller's Albatross
Near Threatened

species (63 per cent) being endemics, most of them to New Zealand. Of these *Vulnerable* endemics, confident long-term assessments of the overall population status are lacking for **Antipodean albatrosses**, a species known to be killed by fisheries operations in both national and distant waters. The population status of **southern royal** and **Campbell albatrosses** have shown increases at some colonies, although their overall status is assumed to be stable. The status of the **Salvin's albatross** population is also assumed to be stable, although lack of comparable population data makes the accuracy of this assessment uncertain.

For the *Vulnerable* **wandering albatross**, the world's second largest population occurs at South Georgia, where a 30 per cent decline has been documented since 1984 (Poncet et al. 2006). Interactions with fisheries have been identified as the most likely cause of this decline, this also being the case for populations of other species breeding in this region. Over half the world's **grey-headed albatrosses** breed at South Georgia, and recent surveys have documented a decline of over 30 per cent since 1990 (Poncet et al. 2006). If these declines are reported for other breeding sites, of which there is no current information, the listing for this species will likely rise to *Critically Endangered*.

The **short-tailed albatross** population is increasing in numbers, albeit a recovery from previous exploitation. Extensive conservation efforts have been undertaken at the volcanic Torishima breeding site of short-tailed albatrosses, the success of which enabled this species to be down-listed from *Endangered* to *Vulnerable* in 2000. **Laysan albatrosses** are the second most abundant species of albatross and were listed as *Vulnerable* for the first time in 2003 as a result of population declines attributed to the effects of longline fishing in the North Pacific. Ingestion of plastics also poses a threat to these species that forage extensively in the Pacific Ocean.

Light-mantled albatrosses have the lowest level of risk of extinction and are currently considered *Near Threatened*. As a result of the precipitous and dispersed nesting localities of this species, population trends remain largely unknown, including the largest populations at South Georgia, Kerguelen and the Auckland Islands. While this species is well known to be killed in fishing operations, it is typically one of the less frequently encountered bycatch species. **White-capped albatrosses** are known to be exposed to fisheries that kill many thousands of albatrosses each year (Baker et al. in press) and information on population trends for this species is currently lacking. Similarly, difficulty in application of robust survey techniques at the largest colony of the Australian endemic **shy albatrosses** confounds an accurate assessment of the overall population trend, although the recovery of the Albatross Island population has been well documented. The steady increase of the **Buller's albatross** at The Snares caused its downlisting to this category in 2008.

This assessment paints a solemn forecast for the survival prospects of albatrosses. How much longer can these birds endure the impacts of the wide-scale threats that we continue to place in their path? The overall trends of half of the 22 species are plunging. The critically small size of some populations, and indeed some entire species, coupled with alarming rates of decline, looks like continuing the legacy of other iconic — and now extinct — marine bird species.

Albatrosses, however, share their stage with a determined band of people across the world who are committed to their conservation. Tides can be turned — as so wonderfully shown by the story of the short-tailed albatrosses breeding on Torishima. It will require the urgent, continued and escalated efforts of all these people, representing industries, governments and non-government organisations, to commit to working collaboratively to improve the imperilled conservation status of this flagship group of birds.

Atlantic Yellow-nosed Albatross
Endangered

Indian Yellow-nosed Albatross
Endangered

Laysan Albatross
Vulnerable

Black-footed Albatross
Endangered

Short-tailed Albatross
Vulnerable

Waved Albatross
Critically Endangered

Light-mantled Albatross
Near Threatened

Sooty Albatross
Endangered

Albatross populations: status and threats

Species and status	Population decline	Restricted breeding range	Limited population size	Decline in habitat	Endemic to single country	Number of sub-populations	Annual breeding pairs*	Annual/biennial breeder	Current population trend**
CRITICALLY ENDANGERED									
Tristan albatross (*Diomedea dabbenena*)	•	•			United Kingdom	1	1500–2400	B	↓
Chatham albatross (*Thalassarche eremita*)		•		•	New Zealand	1	5300	A	↕
Amsterdam albatross (*Diomedea amsterdamensis*)	•	•	•		France	1	18–25	B	↓
Waved albatross (*Phoebastria irrorata*)	•	•			Ecuador	2	15,600–18,200	A	↓
ENDANGERED									
Northern royal albatross (*Diomedea sanfordi*)	•	•		•	New Zealand	3	6500–7000	B	↓
Black-footed albatross (*Phoebastria nigripes*)	•					6	54,500	A	↓
Black-browed albatross (*Thalassarche melanophrys*)	•					7	530,000	A	↓
Atlantic yellow-nosed albatross (*Thalassarche chlororhynchos*)	•				United Kingdom	2	26,600–40,600	A	↓
Indian yellow-nosed albatross (*Thalassarche carteri*)	•					4	36,500	A	↓
Sooty albatross (*Phoebetria fusca*)	•					5	12,500–19,000	B	↓
VULNERABLE									
Wandering albatross (*Diomedea exulans*)	•					5	8500	B	↓
Antipodean albatross (*Diomedea antipodensis*)	?	•			New Zealand	3	11,000	B	?
Southern royal albatross (*Diomedea epomophora*)		•		•	New Zealand	2	8400	B	↕
Salvin's albatross (*Thalassarche salvini*)		•			New Zealand	2	30,750	A	↕
Short-tailed albatross (*Phoebastria albatrus*)		•	•		Japan	2	c. 400[1]	A	↑
Laysan albatross (*Phoebastria immutabilis*)	•					3	437,000	A	↓
Campbell albatross (*Thalassarche impavida*)		•			New Zealand	1	23,500	A	↕
Grey-headed albatross (*Thalassarche chrysostoma*)	•					7	92,300	B	↓
NEAR THREATENED									
Light-mantled albatross (*Phoebetria palpebrata*)	?					6	19,000–24,000	B	?
White-capped albatross (*Thalassarche steadi*)	?	•			New Zealand	2	c. 110,000[2]	A	?
Buller's albatross (*Thalassarche bulleri*)		•			New Zealand	3	32,000	A	↕
Shy albatross (*Thalassarche cauta*)	?	•			Australia	3	12,200	A	?

* Not all pairs breed annually therefore total numbers of breeding pairs remain elusive. A "B" next to the breeding numbers further indicates biennual breeding.
** Key to population trend symbols: ↑ increasing ↕ stable ↓ decreasing.

[1] H. Hasegawa/R. Suryan, pers. comm..
[2] B. Baker pers. comm.

11
Albatross Flight Performance and Energetics
Scott Shaffer

Dr Scott A. Shaffer is a research biologist and lecturer at the University of California Santa Cruz, California, USA. He has been studying the foraging ecology, flight performance, and energetics of albatrosses and petrels in the Southern Ocean and Hawaiian Islands since 1997. shaffer@biology.ucsc.edu

Synopsis: A detailed analysis of how albatrosses achieve what no other bird does: harnessing the power of wind and gravity to fly enormous distances with minimum energy expenditure. The trade-offs that this lifestyle requires are also considered.

I remember distinctly the first time I saw a wandering albatross in flight. I was aboard the RV *Marion Dufresne*, heading south from Reunion Island to the French subantarctic Crozet Archipelago, when a large female wandering albatross approached from a distance. As the bird began to follow the ship, I became mesmerised by the effortless glide and speed of its flight as well as its large body size compared to all the other seabirds flying around the vessel. I imagined sailors, hundreds of years earlier, being fascinated like I was by the magnificent flight of these birds. This experience initiated my curiosity into albatross flight performance and the energetic costs associated with their remarkable flight capability. Together, these facets of albatross biology are fundamental to understanding how these seabirds can raise offspring by commuting to and from food resources located thousands of kilometres from their nest.

Tracking studies reveal that albatrosses can travel 600 kilometres (373 miles) per day on average and reach instantaneous speeds exceeding 100 kilometres per hour (62.5 mph) while engaged in gliding flight. *How is this possible with such little effort?* The size and shape of the wings in relation to body mass are key features that dictate flight performance in albatrosses. Albatrosses have the longest wings compared to all other seabirds. Approaching 1.5 metres (4.5 ft) in length, the single wing of a wandering albatross is nearly three-quarters the height of an average adult human male. Overall, this gives wandering albatrosses, the largest species, an impressive wingspan of 3.1 metres (10 ft) on average (including body width). In comparison, yellow-nosed albatrosses, the smallest albatross species, have a single wing length less than one metre (3 ft) and a combined wingspan of two metres (6.5 ft). Albatross wings are also narrow despite their given lengths compared to other seabirds. This yields a wing with a comparatively low surface area and high aspect ratio, which is the ratio between the wingspan and wing width. Aspect ratios for albatross wings vary between 13.5 and 15.4 on average, which mean that wingspans are 13–15 times longer than average wing widths. This wing shape is similar to that of sailplanes and is ideal for exploiting prevailing winds that blow along the sea surface because albatrosses can glide at a shallow angle to maximise forward movement with a minimum of drop.

The last component of albatross flight performance that affects the flight speed combines wing and body size to produce a measure called wing loading. Wing loading is defined as the amount of weight (in Newtons = mass x gravity) supported by the unit area of a wing. Given that albatrosses are large seabirds (2 to 10 kg [4.4–22 lb]) with comparatively low wing surface areas (0.29 to 0.61 square metres [3.1–6.6 sq ft]), wing loading is exceptionally high. This means that albatrosses must glide faster to remain airborne compared to seabirds with lower wing loadings, which is a major reason why albatrosses occur in regions with strong persistent winds. The heavy wing loading and large body size also allow albatrosses to penetrate strong headwinds to maintain forward flight without flapping at moderate ground speeds.

In order for a bird to soar without flapping its wings, upward lift must exceed the downward force of gravity on its body. By maintaining a shallow downward glide, and using wind to provide the counteracting lift, albatrosses can soar using updraughts generated by the movement of surface winds across waves or by the characteristic 'see-sawing' flight behaviour known as dynamic soaring. Dynamic soaring can be described as a mode of flight where albatrosses fly into headwinds to increase their altitude followed by a downward path moving with the wind to gain momentum to repeat the cycle. This 'see-sawing' flight behaviour is possible because the velocity of winds blowing across the sea surface increase with height off the water due to a boundary layer effect. Therefore, albatrosses can increase their flight speeds and thus gain altitude by flying into headwinds until stall speeds are reached. At this point, birds will turn and descend rapidly with a tailwind to increase their ground speeds. By repeating the cycle with only modest adjustments in

body and wing orientation, albatrosses can achieve forward movement by parasitising wind energy. Furthermore, albatrosses have a special tendon that locks the outstretched wings in place, so little energy is used to keep the wings extended. Essentially, the albatross body plan is perfectly designed to maximise gliding performance at high speeds with a minimum of effort.

However, this type of flight performance comes with a significant trade-off. Though great for gliding, the long narrow wings are poor at producing adequate lift to sustain flapping flight, which is why albatrosses are clumsy fliers with poor manoeuvrability at low flight speeds. This is really noticeable when albatrosses take off and land in their nesting areas. I remember many times seeing a wandering albatross tumble into a somersault upon landing in the colony. I have also seen albatrosses 'stranded' during periods when winds were too light for birds to take off and become airborne.

Flapping flight is not an activity that albatrosses perform for prolonged periods. In fact, previous studies show that taking off and landing are the most energetically expensive behaviours that albatrosses engage in while at sea. Heart rates of wandering albatrosses taking off from the sea surface are nearly triple that when incubating an egg, or about double that of gliding flight. This makes the overall energetic cost for albatrosses to travel around the open sea while searching for food low compared to other seabirds.

In order to compare the energetic costs among animals with different body masses, which affects metabolic output, it is common to derive the ratio between individuals' active and resting states (i.e. active metabolic rate/resting metabolic rate). Using these comparative ratios we find that the energetic cost of a wandering albatross (c. 10 kg or 22 lb) to find food is about two times resting, whereas the cost to a sooty tern (c. 150 g or 5.3 oz) is three times resting. The cost is even higher for a plunge-diving seabird like the northern gannet, which can be as high as six times resting. These comparative ratios — which are among the lowest for vertebrate animals that fly — clearly illustrate the economical nature of soaring flight in albatrosses.

Flying so fast and expending so little energy, an albatross can search for food thousands of kilometres from its nest. For example, a wandering albatross that I studied in the Crozet Archipelago travelled 10,000 kilometres (6200 miles) in 14 days and the energetic cost to conduct this trip was approximately two times the cost of the bird at rest. Truly amazing!

But this foraging strategy comes with another trade-off: low energy delivery to nestlings because adults visit their nests less often compared to other seabirds. Consequently, albatrosses raise only one chick per breeding season and the development time is protracted (7–11 months), such that some species skip breeding in consecutive years. These life-history patterns illustrate the interconnectedness between flight performance, energy expenditure, breeding and the ocean-going habits of albatrosses. In essence, albatrosses are 'built' to take full advantage of wind energy to travel rapidly over the sea surface at a low energetic cost, which makes them the ultimate flying machine.

OPPOSITE TOP By skimming the sea surface on a windless day, a northern royal albatross uses the ground effect to maintain lift.
OPPOSITE BOTTOM A black-browed albatross veers into the wind, its wings briefly reaching a vertical plane.
BELOW Various stages of dynamic soaring can be seen in a flock of black-browed albatrosses rising over a wave crest in the company of smaller sooty shearwaters.

12
Do Wanderers Always Return?
Michael Double

Dr Michael Double is an ecologist and geneticist who works for the Australian Antarctic Division and is a visiting fellow in the School of Botany and Zoology at the Australian National University in Canberra. He is also the Convenor of the Taxonomy Working Group for the Agreement on the Conservation of Albatrosses and Petrels (ACAP). Mike.Double@aad.gov.au

BELOW The northern royal albatross is the only species known to have established a new colony on a site devoid of other albatrosses. Seventy-six years after the first lone pair settled on New Zealand's Taiaroa Head, the colony hatched its 500th chick in 2007.

Synopsis: While most albatross return faithfully to their natal islands to breed, examples of strays and dispersers, and even the rare establishment of new nesting colonies, help shed light on their genetic relationships.

Each April on Albatross Island young shy albatrosses leave their nests and mass on the edge of the island's steep cliffs. There they stretch their long wings and exercise before eventually launching on their maiden flights over the ocean near Tasmania's north-west coast. Hunkered down in the wet tussock, silently watching these apprehensive birds, I wonder how many will survive and how far will they venture, but what also intrigues me is whether any will choose to breed elsewhere.

Adult albatrosses almost always return to the island where they bred previously (behaviour known as breeding philopatry). This was shown by a 40-year study of wandering albatross in the Crozet Islands, in which over 8000 birds were banded but only three adults were recorded as breeding on more than one island (Inchausti and Weimerskirch 2002). Even young albatrosses, after many years at sea and travelling tens of thousands of kilometres, almost always return to breed on the island from where they fledged (natal philopatry). Amazingly, from nearly 900 Buller's albatrosses banded as chicks, of those surviving to adulthood, over half settled within 100 metres (328 ft) of their natal nest and none moved to other islands where the species breeds (Sagar et al. 1998).

This 'homing' ability is remarkable because most albatross breed on remote oceanic islands and must therefore navigate back to these islands in a watery world devoid of obvious 'landmarks'. How they do this is poorly understood, but their navigational skills were exemplified in a bizarre experiment conducted at Midway Atoll in the 1950s. From there, military aircraft flew 18 Laysan albatrosses to different parts of the Pacific. All but four birds returned to their breeding island, including one bird that was released more than 6500 kilometres (4038 miles) away (Kenyon & Rive 1958). More recently, albatrosses carrying transmitters have shown they can even circumnavigate the globe then successfully relocate their breeding islands (Croxall et al. 2005).

Armed with such impressive navigational skills, it is perhaps not surprising that albatrosses favour the island they know best — but inevitably in nature there are exceptions to the rule. Young albatrosses looking to breed for the first time do occasionally nest on a non-natal island, and once established, they are likely to remain faithful to this island throughout their lives. Banding and genetic studies have now revealed that for some species movement between islands can be relatively common. In the Auckland Islands group, white-capped albatross breed on three islands but these populations commonly exchange individuals so the colonies show few genetic differences (Abbott and Double 2003). This is not particularly surprising given these islands are only separated by a few kilometres, but such exchanges can also occur on a much greater scale. Banding and genetic studies of wandering albatross found that some birds fledging from the Crozet Islands have subsequently bred on the Prince Edward Islands and Kerguelen Islands some 900 kilometres (560 miles) to the west, and 1400 kilometres (870 miles) to the east respectively (Inchausti and Weimerskirch 2002, Weimerskirch et al. 1997). Similarly, a wandering albatross from Macquarie Island was found breeding on Heard Island, 5000 kilometres (3000 miles) to the east (Johnstone 1982). Such long-distance dispersal may

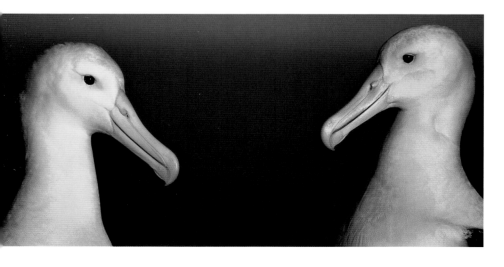

occur in other species too; no genetic differences have been found between any grey-headed albatross populations despite the colonies being widely dispersed throughout the Southern Ocean (Burg and Croxall 2001).

While some albatrosses move to other colonies, others go astray in a rather different way — perhaps through crisis of identity. In 2003 and 2004 a white-capped albatross was sighted among the black-browed albatrosses on South Georgia in the South Atlantic, around 8000 kilometres (4970 miles) from its breeding sites south of New Zealand (Phelan et al. 2004). Records of individuals unexpectedly hanging around in colonies of a different species are, in fact, surprisingly common. For example, Chatham albatrosses have been recorded repeatedly visiting shy albatross colonies off Tasmania (Lindsay 1986) and Salvin's albatross colonies on The Snares (Miskelly et al. 2001). The Salvin's albatross reciprocate by visiting the breeding Chatham albatrosses. Salvin's albatrosses have also been recorded in a black-browed albatross colony in Chile (Arata 2003) and, amazingly, in 2003 one was photographed sitting among the Laysan albatrosses of tropical Midway Atoll (Robertson et al. 2005).

But the most extraordinary records of albatross dispersal must come from the North Atlantic. In 1972, a lone black-browed albatross was seen sitting among breeding gannets on the Shetland Islands, Scotland, some 13,000 kilometres (8000 miles) from the species' nearest breeding colony. Rather sadly, this bird returned to the same spot for the following 25 years (Mead 1996). Remarkably, about a century before, another black-browed albatross did a similar thing, keeping company with the gannets of the Faroe Islands for 34 breeding seasons until it was shot in 1894.

These stories of bizarre navigation and mistaken identity are perhaps rather tragic but at the same time fascinating. I suspect these mishaps may tell us more about the distribution and evolution of albatross. What if birds of the opposite sex but the same species made the same mistake simultaneously? As once speculated (Mead 1996), what if the two black-browed albatrosses of the North Atlantic had not arrived a century apart; would there now be a colony of albatrosses in Scotland?

The intriguing distribution of some albatross species hints that, given enough time, these coincidences must occur and new breeding colonies form. In species such as the sooty (Kerguelen Island), wandering (Heard Island), Antipodean (Campbell Island), and royal (Taiaroa Head) albatrosses, colonies of fewer than 10 pairs have been documented well away from their main population centres. But perhaps the most amazing cases are those of Salvin's and Indian yellow-nosed albatrosses. Over 80,000 pairs of Salvin's nest on the Bounty and Snares Islands south of New Zealand yet a lonely group of four pairs nest over 8000 kilometres (5000 miles) away in the middle of the Indian Ocean on Île des Pingouins in the Crozet Islands (Jouventin 1990). This situation is reversed for a single intrepid pair of Indian yellow-nosed albatross nesting alone among the Chatham albatrosses of The Pyramid at least 7000 kilometres (1242 miles) distant from the remaining 36,000 pairs breeding on several islands in the Indian Ocean (Miskelly et al. 2006).

What is not always clear, though, is whether these stragglers are remnants of a once larger population or newly established dispersers. In only a few cases is it relatively obvious that the colonising species was new to a site because the area was well known. As far as I know there is only a single well-documented case of albatrosses colonising a site otherwise devoid of other albatross species. That is the famous case of the royal albatrosses of Taiaroa Head near Dunedin, New Zealand. There, in 1920, a single pair of albatross nested on the bare headland for the first time. Within four years there were at least 10 birds nesting so other immigrants must have joined the pioneering pair. This site is now one of the most accessible albatross colonies anywhere in the world and nearly 30 pairs breed there regularly.

Both chance observations and detailed studies have revealed that very occasionally young birds will indeed nest away from their natal island, settling among albatrosses of another species, or perhaps even where no albatrosses have bred before. But it is clear that the formation of an entirely new albatross breeding site is an exceedingly rare event, perhaps occurring less than once a millennium. To know this highlights that places where albatrosses breed today are extremely precious and must be properly managed, protected and cherished.

FAR LEFT In January 2008 two juvenile short-tailed albatrosses landed on Hawaii's Midway Atoll, one of them soon joining the lone adult keeping company with the 42 decoys on Eastern Island (the smaller of the two in the atoll) placed there in hopes of creating a new breeding colony. From their size difference, the adult appears to be a male and the newcomer a female. The male was banded as a fledgling in 1988 on Torishima near Japan, and has been coming here each season since 1999. An adult female (described in Part I) also visited from the mid 1990s to 2004, but the two never met ashore as they frequented separate parts of the atoll.

BELOW Having strayed into the wrong hemisphere, for 22 years — from 1973 to 1995 — a lone black-browed albatross, seen here in May 1991, visited the large gannet colony at Hermaness in the northern Shetland Islands of Scotland.

13
Albatross Populations and Migrations: From Observation to Application
John P. Croxall

Prof. John P. Croxall, CBE is currently Chair of BirdLife International's Global Seabird Programme. From 1976 to 2006 he was in charge of bird and mammal research for the British Antarctic Survey, latterly as Head of Conservation Biology. He was Chairman of the Royal Society for the Protection of Birds 1998–2003, appointed CBE for services to ornithology and marine conservation in 2004 and elected Fellow of the Royal Society in 2005.

Synopsis: Long-term population studies of three species of albatross from South Georgia reveal major declines over 30 years which, together with other regional studies, were key to uncovering alarming global mortality due to high-seas fisheries. Long-distance tracking enables risks of fisheries-related bycatch to be assessed and remedies applied.

When Peter Prince and I started systematic population studies of wandering, grey-headed and black-browed albatross at Bird Island in the early 1970s, we were privileged to inherit many known-age birds ringed by Lance Tickell and Ron Pinder between 1958 and 1964, and to be spoilt for choice in selecting colonies and areas for study. Little did we know that 30 or so years later the survivors of these birds and their offspring would be witness to and victims of the best documented of all the albatross population declines and that many of the birds (and indeed colonies) that we studied would have disappeared. What began as a study of relationships between age, experience, performance and survivorship came to produce data crucial for ensuring survival of albatross populations and even species.

As early as 1979 we suggested that wandering albatross breeding numbers at South Georgia might be in decline, based on the accurate annual counts from Bird Island and some supporting data from aerial photography of birds in the Bay of Isles. There were, of course, a number of possible causes to investigate. We were able quite quickly to rule out disturbance by scientists and by Antarctic fur seals. To investigate the first we had several large areas and colonies visited only for annual counts and otherwise undisturbed, yet these showed the same trends as the more intensively studied areas and colonies. In contrast, fur seals had certainly displaced breeding wandering albatross from most low-lying parts of Bird Island. However, detailed studies showed that seal disturbance caused an initial local reduction in breeding success and subsequent redistribution of adults (because young recruiting birds avoided areas with seals) but did not reduce overall numbers.

By the late 1980s mounting evidence was accumulating of a pattern of deaths associated with longline fishing for tuna in the waters off southern Brazil, Uruguay and northern Argentina. By 1990 we were able to demonstrate statistically significant population declines (not easy given the inter-annual variation inherent in biennially breeding albatross species such as Wanderers) linked to a significant adult mortality. The commonest cause of deaths was in the longline fisheries and we surmised that this was the problem. At the same time Nigel Brothers, working in Australian waters, revealed that annual mortality levels in Japanese tuna fisheries were in the order of 40,000 albatrosses. Further research confirmed and extended these findings, with models of tuna longline fishing effort matching the declines of Atlantic and Indian Ocean wandering albatross populations (together representing over half of the world population).

By the mid-1990s, declines of similar magnitude were evident in South Georgia populations of grey-headed albatross (at the world's premier breeding site) and black-browed albatross (after the Falklands, the main world site). Since then Bird Island populations of all species have continued to decline — at increased rates for black-browed and wandering albatrosses — and are nowadays less than half the size of when we started our research. The 2003/04 complete census of South Georgia albatrosses confirmed that similar declines are occurring island-wide. None of these species currently faces any problem at their breeding colonies so their fate is inextricably linked to how we manage threatening processes, especially fishing, in the southern oceans.

One of the reasons South Georgia albatross populations may be more vulnerable than many others could relate to their migration patterns. In the early stages of albatross research at Bird Island, migration destinations were discovered through the recovery of banded individuals. Thus by the 1970s it was well known from work by Lance Tickell at Bird Island and Doug Gibson in New South Wales that there was regular, sometimes annual, movements of wandering albatross between the two areas — the New South Wales 'hot spot' promoted by local discharge of untreated sewage! It was also known that many South Georgia black-browed albatrosses moved to

ABOVE RIGHT A grey-headed albatross delivering food to its chick is likely to have risked dangerous interactions with industrial fishing operations on many of its prime feeding grounds, where bycatch remains a serious threat.

South Africa in winter. Recovery of bands from birds killed on fishing boats in the 1980s also showed that wandering and grey-headed albatrosses from Bird Island regularly ranged north to 30 degrees south — well within the range of tuna fisheries.

However, the start of the real breakthrough in understanding the at-sea ranges of albatrosses came in 1990 when the first satellite tracks from wandering albatrosses instrumented with PTTs (satellite transmitters) became available. Data from Bird Island studies showed that while rearing chicks they regularly travelled to the waters off southern Brazil, on round trips of 10,000–12,000 kilometres (6200–7450 miles), thus running the gauntlet of the longline fishing fleets operating in these areas. Subsequent work accidentally (because we failed to recover some devices!) showed that, after breeding, our wandering albatrosses took only seven days to travel to southern Africa but another 15 days to reach Australia. More recently, even these journeys have been put into perspective by the amazing travels of grey-headed albatrosses, studies facilitated by the British Antarctic Survey's development of tiny light detection loggers (geo-locators) capable of providing two positions per day for several years. This showed that not only could breeding grey-headed albatrosses circumnavigate the globe between successive breeding attempts, but that many individuals did so twice. Extraordinarily, they frequented similar areas on both traverses. The fastest bird made a complete circumnavigation in 46 days; all birds used areas of important albatross habitat south-east of South Africa, avoided substantial areas of apparently suitable habitat in the eastern Indian Ocean and spent appreciable time in particular areas in the Pacific. We were able to show that although the Bird Island population employs at least three distinct migration strategies, individuals consistently exploited the same staging areas in successive winters. Furthermore, despite the apparent flexibility offered by their lengthy non-breeding season, the timing and duration of each leg of the migratory journeys was well synchronised between individuals and within strategies. In addition, all individuals used the breeding season home range around South Georgia for most of the non-breeding summer between two successive breeding attempts.

In contrast to wandering albatrosses from the Indian Ocean and many of the species in the Australasian region, grey-headed and wandering albatrosses from South Georgia regularly travel across all ocean sectors south of 30 degrees south and thus frequent most of the areas where longline fishing takes place. They are at potential risk of mortality for a much greater proportion of their at-sea life than most other species and populations. In addition, South Georgia black-browed albatrosses are now known to frequent the numerous longline and trawl fisheries operating in the Benguela and Agulhas Currents around southern Africa, where the risk of incidental mortality is among the highest currently documented.

Our data and approaches to defining foraging range were central to the recent BirdLife International initiative combining remote-tracking data for albatrosses and petrels to provide a global overview of their ranges. In particular, this enabled the first assessments of the key areas where these species overlap with fisheries in which albatross bycatch is a known or potential issue. The resulting publication, *Tracking Ocean Wanderers* (BirdLife 2005), has created global interest and, together with BirdLife's review of the performance of Regional Fishery Management Organisations (Small 2005), has enabled involvement with many of the key fishery organisations in seeking to start new initiatives to reduce levels of bycatch albatrosses and petrels worldwide.

These two case histories indicate how research, which started as a purely intellectual endeavour, has come to be central to addressing fundamental issues of management of marine systems which hold the key to the future survival of albatross populations.

ABOVE LEFT AND RIGHT A grey-headed albatross colony at Cape Alexandra, South Georgia photographed in 2000 (left) and 2005 (right) shows visibly declining numbers, a trend seen throughout the region. Likewise, most wandering and black-browed colonies have shrunk by about 30 per cent since the mid-1980s.

BELOW Three non-breeding grey-headed albatross satellite tracks from Bird Island near South Georgia show complete circumnavigations of the earth, the fastest doing so in just 46 days.

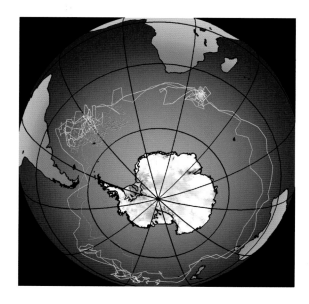

14
Indian Ocean Albatrosses: A Status Report
Henri Weimerskirch

Dr Henri Weimerskirch is Director of Research in Centre National de Recherches Scientifiques, and head of the Marine Top Predators Group at the CEB of Chizé, France. He has worked extensively on oceanic islands species, especially in the Indian Ocean.
henriw@cebc.cnrs.fr

BELOW Avian cholera, possibly introduced to Amsterdam Island through human activity, has caused massive mortalities among Indian yellow-nosed chicks, resulting in the near failure of recent breeding seasons.

Synopsis: Analysing population trends of the Indian Ocean albatross species raises questions about what could be the driving factors for declines.

The French Southern Territories (Terres Australes et Antarctiques Françaises — TAAF) include three groups of islands in the Southern Indian Ocean: the Crozet Islands (C), the Kerguelen Islands (K) and Amsterdam and Saint Paul (A/SP). These islands have been the centre for research on albatross ecology over the past 45 years, beginning with the works of Mougin in the 1960s. The earliest population numbers date from 1959–60 when Tilman visited the Crozets to band breeding wandering albatrosses and counted the birds spread over the eastern part of Possession Island. Since 1964, the continuous presence of an ornithologist has allowed long-term recording of the island's populations. At Kerguelen and Amsterdam, studies on albatrosses started in the early 1970s, with continuous monitoring since 1979. The other islands with albatrosses in this region are the Prince Edward Islands (PE) and the MacDonald-Heard Islands (MDH), administered by South Africa and Australia respectively.

Eight species of albatrosses breed in the Southern Indian Ocean: wandering (C, K, PE), Amsterdam (A), black-browed (C, K, MDH), grey-headed (PE, C, K), Indian yellow-nosed (PE, C, K, A/SP), Salvin's (C), sooty (PE, C, K, A/SP) and light-mantled (PE, C, K, MDH). The trends of the different species, except for the few Salvin's, are known from the long-term monitoring studies carried out in the French Territories. Since continuous monitoring began in the 1960s, when the population appeared stable, the wandering albatross population dropped by half in the 1970s, then increased from the mid-1980s until 2000, when it started to decrease again. Earlier studies (Weimerskirch and Jouventin 1987) had suspected that the decline could be related to fisheries since some birds were found injured with hooks. The potential effect of fishing practices on albatrosses was later confirmed when Brothers (1991) indicated that probably 10,000 albatrosses were killed annually by longline fisheries, especially in the Indian Ocean southern blue-fin tuna fishery that started to expand in the early 1970s. Since these reports were published, there has been an extensive effort in the Southern Ocean, and later worldwide, to estimate the status of albatrosses and other Procellariiformes and their susceptibility to longline fisheries. Several accounts have provided good evidence that albatross populations are declining at some sites in the Atlantic, Indian and Pacific Oceans, and in most cases longline fisheries were suspected to be the cause. While there is little doubt that such fisheries have impacted on many albatross populations at many sites, there has been a strong tendency to implicate these practices without substantiating evidence. The long-term studies in the French Territories are providing some suggestions that certain population changes are more probably influenced by a complex set of environmental as well as human-induced parameters.

Wandering albatrosses are by far the best-documented case, through long-term studies carried out at South Georgia, Prince Edward, Crozet and Macquarie. Although all studies concluded that population declines were due to longline fishing, much remains unexplained in the patterns observed if this fishery alone was the main factor, as it is in the case of the continuously declining South Georgia population (Tuck et al. 2003). The fact that the patterns observed for the Crozet population were later found to be similar at both Prince Edward and Kerguelen islands indicates that common factors were influencing Indian Ocean wandering albatrosses, whereas fishing effort has been extremely different in the foraging zones of each population. Other factors, such as large-scale environmental changes, could have been a confounding factor affecting albatross populations as well as other squid-eating species that have declined at the same time as wandering albatrosses (Weimerskirch et al. 2003).

Amsterdam albatrosses are a subtropical equivalent of the subantarctic wandering albatrosses. In the early 1980s the population consisted of just five nesting pairs per year, and there is no indication as to why their numbers were so low at that time. Since then, they have increased steadily at an incredible rate (for an albatross), to

attain 30 nesting pairs in 2006. Amsterdam Island lies close to the zone of heavy longline fishing and the ability of the population to increase suggests that currently this activity has no influence on the population. On the other hand, the sooty albatross from Crozet, another subtropical species, has declined continuously over the past 25 years to reach one third of its original numbers. Sooty albatrosses are rarely taken as bycatch by longliners, with only a single individual recovered. Conversely, the closely related subantarctic light-mantled albatross from Crozet has shown no significant changes over the same period.

Similarly, Kerguelen black-browed albatrosses demonstrate no significant long-term trends, although their numbers have varied cyclically. A demographic analysis indicates that adult survival is negatively influenced by longline fishing in their wintering zones, but positively so by trawling around the breeding zone. Breeding success and juvenile survival is also enhanced by warm conditions around the breeding grounds. The net outcome is a fluctuating population with an overall stability. Cycles in population size are directly related to cyclical patterns in the Southern Ocean, with a lag due to recruitment. Therefore in this region the longline fishery has a reduced influence that does not affect the overall stability of the population.

Indian yellow-nosed albatrosses at Amsterdam declined steadily in the 1980s. This was originally suspected to be the result of bycatch, but the numbers caught in the wintering zones off Australia were not sufficient to explain the collapse at Amsterdam. Additional mortality around the breeding grounds was a possible compounding factor, although no band recovery was available for this area. It was then found that a worldwide disease, avian cholera, was affecting the population, killing chicks in large numbers. The pathogen started to spread in the early 1980s. Its origin is not known, but it could have spread from poultry at the Amsterdam Island base.

These examples from the Indian Ocean indicate that demographic parameters driving population numbers can be affected by a series of factors. Since albatrosses are long-lived, the sensitivity of the population is influenced by factors primarily affecting adult survival, secondarily immature survival, then fecundity. Natural selection has probably resulted in adult survival having developed a high tolerance to environmental stochasticity (a sequence of random variables). If a factor such as longline fishing directly impacts adult survival, it is likely to have severe effects on the population. However, other causes, such as diseases, or deep environmental changes such as oceanic regime shifts, can also affect adult as well as juvenile survival and fecundity, thus can be major elements driving the population changes. For example, it has only recently been recognised that a regime shift has probably affected the Southern Ocean in the Atlantic and Indian sectors, and that krill stocks have probably decreased by an order of magnitude (Atkinson, Siegel, Pakhomov and Rothery 2004). Such changes will have undoubtedly affected many albatross species and it will be important in the future to use not only fishing effort, but also environmental parameters as variables to explain population changes.

15
Waved Albatross
David Anderson

Dr David Anderson is Professor of Biology at Wake Forest University in North Carolina, USA. Most of his research has focused on the ecology, evolution and behaviour of the seabirds of the Galapagos Islands. da@wfu.edu

Synopsis: A personal account of 25 years studying one of the most unusual of all albatrosses, investigating their habits, movements and threats, and also some enduring mysteries, such as why most incubating adults move their eggs well away from where they were laid.

On a foggy day, Española Island, in the south-east corner of the Galapagos Islands, could pass for J.R.R. Tolkien's Mordor. I first walked along the coast of this broken heap of volcanic debris in 1981, twisting and crawling to pass through the jagged grey vegetation, spooked by the gnarled black marine iguanas scuttling every which way. Rounding the eastern point at Punta Cevallos, climbing over yet more bubbly and blasted rock, a starker contrast could hardly have presented itself to Frodo: scores of waved albatrosses between the lava boulders, their elegant bright yellow bills somehow fracturing the brutality of the ashen landscape.

These birds immediately impress you with their unassuming countenance as they take in their world, and with their simultaneously delicate and furious courtship dances. My colleagues and I have returned

TOP RIGHT Rough and tumble matings, along with fights, frequently happen near the landing areas at the beginning of the nesting season.
ABOVE RIGHT Rather than build a nest, a waved albatross may move several dozen metres while incubating, holding the egg clamped between its tarsi.
RIGHT Satellite tracks obtained between 1995 and 2005, reveal the commuting routes of waved albatrosses nesting on Española Island and feeding in the Humboldt Current upwellings along the Peruvian coast, recording regularly travelled distances of 3000 kilometres (1864 miles) or more.

nearly every year since then to add chapters to our understanding of their biology, and the story is a long one. The dreamy looks and neck nibbling of pairs belie the seriousness of their breeding intentions. Fights are frequent and I once watched two males with 'a history' tear into each other for 14 hours, leaving both bloody and one without a right eye. Bird 013 was never seen again after that day. Females cheat on their mates, and males run down unwary females and force copulations on them: as a consequence, males frequently raise a chick that is not their own. The foraging trips of waved albatross are phenomenal. While residents of Galapagos, most of their feeding occurs off the Peruvian coast, 1700 kilometres (1056 miles) away. Leaving the safety of the Galapagos Marine Reserve, they risk death in nets and on hooks of fisheries there, and both annual survival and population size appear to have declined abruptly in recent years. To find out what happens at the other end of their commute, we have started working with Peruvian researchers to place observers on board small fishing boats operating off the coast of northern Peru. Out of 22 such trips in our first season 12 waved albatrosses were killed, nine of them by a single boat. Some drown accidentally in fishing gear, but others are caught intentionally as human food, which could have disastrous effects for such a restricted-range species.

A limited, tropical distribution is not the only unique feature of the species, as some of its habits too are baffling. One such research nut we have been unable to crack, is that alone among birds, waved albatrosses regularly leave what passes for a nest site, taking their egg with them. Most albatross species use a volcano-like structure which they build to house the egg and later the chick, and birds in general use nests with varying degrees of enclosure, but no variation in occupancy. The eggs go in the nest, and the eggs stay in the nest, but with waved albatross, 93 per cent of eggs are moved out. The incubating parent scoots the egg along between its tarsi (lower leg bones), and remarkably does not often break the egg during the movement itself. But this can get the egg into places like holes or under bushes that stymie further movement, or even effective incubation. The more frequently an egg is moved, and the further the distance, the less likely that egg will hatch. No other bird regularly moves its eggs (except emperor and king penguins who carry them atop their feet under a flap of warm skin), probably because it's risky, and it would appear that waved albatross should not move theirs either: losses attributable to egg movement and resulting egg marooning account for up to 80 per cent of all breeding failure.

Some hypotheses have been imaginative, focusing on particular problems that might lead to this special behaviour. For example, ground-nesting birds typically do not have to deal with giant tortoises, and Galapagos is full of tortoises. Tortoises might plough ahead like a bulldozer, crushing an egg that parents don't move. This idea could account for the initial evolution of egg-moving behaviour, but fails to explain any of the movement that we have documented, because no tortoises were present during our observations of egg movement. Sea lions use some of the same habitat that the albatrosses do, but egg movement occurs whether sea lions are present or not. In fact, movements are greatest in inland areas where the density of albatrosses, sea lions and all other large vertebrates is lowest.

Española is tropical, and albatrosses tend not to be tropical species. Their laboured panting shows that heat dissipation is clearly a problem for incubating birds, so maybe they move their eggs to improve their thermal microclimate. If so, they should move more towards any local shady vegetation than away from it, but in fact they move randomly with respect to surrounding shrubs. They also move less, not more, during the hot day than during the cool night. For a time we were optimistic that egg movement might be a strategy to get eggs adopted, because they often go from areas of low albatross density to areas of high density, and abandoned eggs are sometimes 'babysat' by non-breeding adults. In fact, 18 per cent of babysat eggs eventually hatched, though usually back under the actual parents who, by then, had returned to the egg. Babysitters might bridge the gap of a few days caused by the parents' temporary abandonment, but in fact abandoned eggs that were babysat did not have a higher hatching success than similar eggs that were not babysat. Why an unemployed adult would care to incubate an abandoned egg is another good question. Information regarding population genetic structure suggests that the egg is unlikely to contain a close kin.

Incubating birds are bothered in some years by mosquitoes and in some places by ticks. They might be trying to escape those irritants, but the evidence doesn't support this idea either. We have obtained one helpful insight: individuals that shift their laying site a lot from one year to another also move their egg a lot after laying it. It seems that some birds are habitually dissatisfied with their nest site, and others are less so. Nonetheless, nearly all eggs are moved to some degree. Even after 25 years, we have been unable to identify a reason for so many birds shuffling across the breeding colony, moving their eggs as much as 36 metres (118 ft) across hazardous terrain. Can the nesting habitat on Española really be so universally unsuitable?

BELOW The waved albatross was uplisted to Critically Endangered status in May 2007, based on evidence of high adult mortality, at least in part due to fishing, with average life expectancy abruptly reduced from 20.3 to just 12 years. Working closely with coastal fishermen in Peru, Jeffrey Mangel and Joanna Alfaro of Pro Delphinus (a local NGO), recovered a total of 105 bands — some still attached to legs — from birds that died in artisanal fishing operations. Twenty-three of these bands came from 2,550 birds banded at the author's study site on Española, suggesting roughly one percent annual mortality. Considering that many more bands are never retrieved, the problem clearly is huge. The author's portrait on the opposite page shows him in Lima, following the 2007 ACAP workshop to address this issue, reunited with the remains of old friends he banded on Española in May 1999.

16
Short-tailed Albatross: Dandies of the Deep
Greg Balogh

Dr Greg Balogh is an Endangered Species Biologist with the US Fish and Wildlife Service in Anchorage, Alaska, where he has worked with birds since the mid 1980s. He heads the Short-tailed Albatross Recovery Team, and has spent many months at sea and on an active volcano in Japan working towards the species' conservation. Greg_Balogh@fws.gov

Synopsis: Hands-on, at-sea description of how to boost the recovery of one of the rarest of all albatross, and what is being learned about this success story.

The air was filled with the pungent smell of hundreds of northern fulmars just off the stern of the *Kema Sue*, an 80-foot longliner out of Kodiak, Alaska. We were adrift south of Amlia Island in the Aleutians. The groans, squeals and clucks of excited seabirds fighting for bits of squid were both dramatic and comical. Black-footed and Laysan albatrosses dominated the surface squid wars, but the sporty fulmars were outmanoeuvring the luxury-sized birds to get their fair share. Shy, non-confrontational short-tailed shearwaters ventured under water, zipping beneath the surface flock, snaffling up the stray tentacles that sank, unnoticed, out of reach of the aggressive birds above. Though we were on a commercial fishing boat, we were not fishing. Rather, we were chucking chopped bait over the rail, hosting the best damned bird party in Alaska.

Our real objective was to capture short-tailed albatrosses and fit them with satellite transmitters, or PTTs. When taped to the birds' back feathers, these PTTs would allow us to follow their movements through late summer and early autumn, until the feathers moulted off, dumping the expensive payload into the briny deep. On this day, we had three short-tailed albatrosses attending our bird party. They were the best-dressed birds of the lot; true dandies with golden hoods set off by a bubblegum-pink bill,

the bill's base rimmed by a fine line of India ink. Their black epaulettes on snowy white wings gave the impression of military rank. They outweighed every other bird at the party by a long shot, yet they behaved like wallflowers, remaining far back from the bait-filled dance floor immediately astern. Until they took part in our pelagic rave, they remained out of range of our hand-thrown hoop nets; safe from science.

We kicked the party into high gear when we deployed our secret weapon: a string of stinky black-cod heads and livers. It was like blasting open a *piñata* at a children's birthday party. The birds went nuts for the rich fatty heads. The youngest, boldest short-tailed albatross threw caution to the wind and joined the party. Rob Suryan, Professor of Marine Ecology at Oregon State University, and I checked and rechecked our 1.5-metre (5-ft) diameter hoop nets, making certain they were ready to be flung overboard. If we figure out the cost of the project versus the number of shots these shy birds would give us, we're looking at about $10,000 per throw. Rob exhibits the calm of a major league pitcher in a pennant race.

As he prepares to throw, Rob simultaneously takes into account cross wind, a heaving 2-metre (6-ft) swell, and the target bird's behaviour. Designated bait flingers Karen Fischer and Dan Roby, also of Oregon State University, keep the party going by flipping squid and manoeuvring the cod heads into position for the $10K toss. A misguided tentacle clings to Rob's cheek as he watches the short-tailed albatross out of the corner of his eye. We can't even make eye contact with these buggers without sending them swimming for the horizon, such is their level of wariness. The moment the young shortie is distracted by a brief discussion over bait ownership with a black-foot, Rob springs up and flings his hoop. Both albatrosses are netted and quickly brought on board. The black-foot is immediately released, but the short-tailed albatross is swept into our makeshift operating chamber on the hatch of the ship's cargo hold. Rob has earned his major league paycheck.

We're uncertain why the short-tailed albatross is the most cautious of the seabirds encountered on the commercial fishing grounds of Alaska. We can only assume the behaviour is linked to the species' history

ABOVE RIGHT A short-tailed albatross on its Alaskan feeding grounds easily displaces the smaller Laysan and black-footed albatrosses amid flocks of Arctic fulmars.
RIGHT On a calm day in the Bering Sea a researcher prepares to toss a hoop net to capture short-tailed albatrosses for remote tracking.

of near extinction at the hands of market hunters in the late 1800s. Perhaps it was only the select few, unusually cautious birds that survived this massive slaughter to pass along their unusually cautious genes.

In Hawaii, breeding Laysan albatrosses regard human intruders with indifference, and black-footed albatrosses everywhere try to extract an ounce of flesh from anyone who wanders within beak distance. But short-tailed albatrosses carry their newfound timidity with them even to their breeding colony. Their wariness, however, does not protect them from their biggest threats.

Unlike many southern hemisphere albatross species, commercial fishing is not the biggest threat to the shortie's continued existence. Ninety per cent of short-tailed albatrosses hail from Tsubame-zaki colony on Torishima Island. The colony is perched precariously on an eroding flank of an active volcano. It sits on a steeply sloped fluvial outwash plain formed from torrents of storm water and ash. The water/ash slurry gushes through a small notch in the wall of the volcano's caldera 100 metres (328 ft) above. An ill-timed typhoon can send a flash flood that takes out albatross nests and washes chicks to their deaths on the rocks below. But even this storm-water scourge is not the biggest threat to the species. Ironically, it is the very volcano that killed 125 feather hunters in 1902 that now looms like a spectre of death for the birds themselves.

While an eruption will certainly not cause species extinction like the feather hunters nearly did, a particularly ill-timed blast could knock the population down by over 40 per cent. An international Short-tailed Albatross Recovery Team has determined that setting up new colonies away from Tsubame-zaki's steep slope and off Torishima Island are the most important keys to ensuring the recovery of the species. Japan's Yamashina Institute has recently enjoyed great success in luring breeding short-tailed albatrosses from the main colony to the other side of the island, where gently sloping, vegetated terrain will no doubt enhance survival rates of chicks. Among the 60 life-like albatross decoys sprinkled across the hillside, at least 13 chicks fledged from this colony in 2006, up from four in 2005 and one in 2004. In 2007, a record 24 pairs laid eggs here. Efforts are also under way to try experimental translocations of albatross chicks from Torishima to the Ogasawara Islands, a non-volcanic archipelago to the south.

Back in Alaska, where commercial fishing poses some risk to the species, we are learning all we can about the distribution of the birds through satellite telemetry. Over the course of the past five years, we've learned that short-tailed albatross forage extensively in the very same waters targeted by much of Alaska's commercial fishing fleet, namely the productive continental shelf break areas of the Bering Sea and Aleutian Chain. We've also learned that the younger birds cover a lot more ground than the breeding adults, with individuals wandering through Japanese,

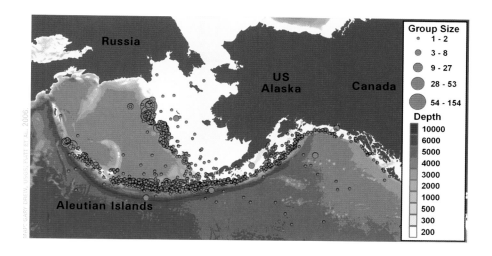

Russian, Alaskan and Canadian waters. In February 2006, we discovered that adults get food for their chicks just off the coastline near Tokyo, also a heavily fished region. But even before we began mapping where the birds wandered, thanks to the precarious status of the short-tailed albatross, scientists and fishermen in Alaska were busy figuring out very effective ways to avoid killing or harming albatrosses in their gear. The resulting drop in seabird bycatch there has been dramatic. But much remains to be learned about the habits of these rare birds at sea. So as Rob and Dan gently release their newly tagged short-tailed albatross overboard, the bird's position immediately begins being recorded from space-based satellites. We move one step closer to unravelling another of the many secrets this species still keeps — secrets that were very nearly lost forever.

ABOVE At-sea short-tailed albatross observation data from the US Fish and Wildlife Service and US Geological Survey (1940–2004) shows a predilection for the continental shelf edge in the Gulf of Alaska and the Aleutian chain, as well as the shelf break in the Bering Sea, where the largest groups were sighted.

BELOW The satellite tracks of 10 non-breeding short-tailed albatross in two age groups, captured in Alaskan waters, show them travelling across the North Pacific, far from their breeding islands off Japan.

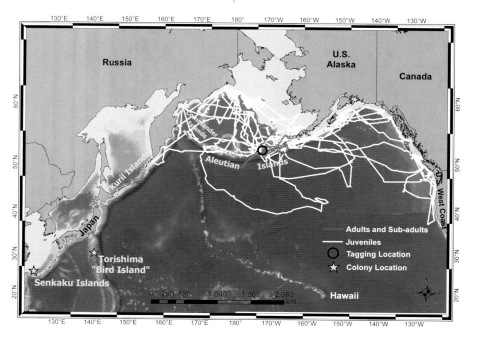

17
Circumpolar Royal Travellers
Christopher J.R. Robertson

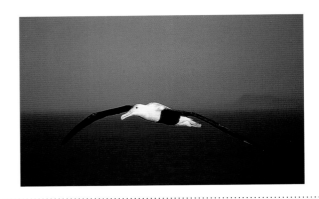

*Dr Christopher Robertson is a past President of the Ornithological Society of New Zealand, and has explored the life of large seabirds and albatrosses over a period of 50 years, from remote storm-swept islands to the bowels of museums worldwide, and, more recently, among the entrails of bycatch specimens on the autopsy table.
cjrr@wildpress.org*

Synopsis: A close look at the full array of portable recording technology reveals the extraordinary travelling prowess of the northern royal albatross, stretching not only our imagination but outpacing satellite software.

During the early 1970s when I was first researching the northern royal albatross on Little Sister Island in the Chatham Islands group off New Zealand, a colleague from Australia, David Nicholls, asked me to trial a package the size of half a tennis ball to carry a transmitter for tracking the albatross. This was pioneering stuff, and I recall telling him that it was too large and cumbersome for the birds to wear. Some more experiments then with various harnesses and smaller packages led to my conclusion that such transmissions still seemed a pipe dream for the future.

By the 1990s technology had advanced and the French Argos weather satellite system had made a special transmission channel available for animal tracking. Transmitters (PTTs) were steadily being reduced in size so that the largest albatrosses were now able to carry a package equal to about three per cent of their body weight. However, it was expensive study, with transmitters costing US$2,000–4,000 each, and satellite time up to US$5,000 per annum per transmitter.

David undertook some of his early tests on the wandering albatrosses that visit the seas off Sydney, Australia each year from the Indian Ocean. However, he wanted to put packages on birds for short periods with some hope of their recovery, so we needed to use birds based at a breeding colony. The first such trials were from the small Taiaroa Head royal albatross colony, near Dunedin, New Zealand, in November 1993, where two short-term deployments were used successfully to test a barometric recorder — albatrosses as flying weather stations in the future perhaps?

During 1994 and 1995 I was again at Little Sister Island. As part of our breeding studies various PTT attachment methods were tested while trialling new transmitters with different regimes to try to extend the life of the batteries, which so far were the main restriction on the duration of experiments. As the larger and better-funded programmes overseas were concentrating on albatross studies during the breeding season, we decided to experiment with turning on the transmitters for only short periods, thus trying to extend the time range towards one year and allow tracking of birds away from the nesting season. From earlier leg-band recoveries we knew that northern royals went to South America — how did they get there and come home we asked?

1 February 1996 saw the first of these specially prepared transmitters depart on 'Agent Orange' (named for his painted orange marker stripe from head to tail), a recently failed breeder. He did not behave altogether as expected, staying in New Zealand waters for much of the year and then departing for South America in September as if he had just raised a chick successfully. He was tracked to Cape Horn before the battery died after 191 days.

Unexpectedly, the female of a pair being trialled at the same time for local foraging movements lost its chick, left the colony and seemingly vanished. With such a tight budget David was somewhat distraught, but Argos kept getting garbled messages that indicated the bird might still be operating, though going too far too fast! The Argos system software applies a test

ABOVE RIGHT Far-ranging research has revealed the extraordinary travels of northern royal albatrosses, such as this one arriving at its nesting grounds on the tiny Sisters islands in the Chatham Islands, New Zealand.
RIGHT Round the world satellite tracking locations for northern royal albatross from Taiaroa Head and Chatham Islands, New Zealand.

to the distance and speed calculations worked out from the positional fixes to assess the validity of the record, which couldn't keep up with our intermittent transmission regime. I recall commenting to David that perhaps 'a New Zealand bird was too fast for a French satellite'. The software was then adjusted by Argos, and the bird was duly 'found' and tracked to the Patagonian Shelf off Argentina until the battery died after 139 days.

We were now ready to try for the 'Round-the-World' challenge! Two failed breeders were harnessed at Little Sister Island on 28 January 1997 and set off with the PTTs transmitting for nine hours and then in sleep mode for 135 hours. Both birds were tracked for 520–560 days (they came home and attempted to breed and failed again). During this time they successfully travelled 1.75 times round the southern hemisphere before the batteries died, with the male taking a generally more northerly (and longer) route than the female. *This was the first tracking of an animal round the southern hemisphere, to and from the Albatross Capital of the World!*

At the International Ornithological Congress in Durban, South Africa in August 1998, I visited the Argos display stand and was startled to find them showing an anonymous, but live, track of one of our birds that was just passing below the Cape of Good Hope. Remarkably, on this second time round, the male bird was no more than 50 nautical miles from its position on the same calendar day, 12 months earlier.

Clearly these birds know where they want to go, and do so very rapidly when on migration — one doing a staggering minimum of 1800 kilometres (1120 miles) in one day — with point-to-point speeds of 80–110 kilometres per hour (50–69 mph) being a regular occurrence. The Argos system, however, can only give positions about every one to two hours so that the fine detail of their foraging patterns had to wait until the arrival of the GPS tracking logger (with positions every second) into the researchers' armoury.

Neither the PTT nor the GPS positional results can directly tell the story of what the bird is actually doing at that point on the globe. There are now stomach temperature recorders that can record when the bird swallows a meal — but then you have to recover the device from the bird.

The cheapest modern tracking tool is the geo-location logger that allows the calculation of one position a day from changes in the time of dawn and dusk as the bird travels. When combined with water temperature logs, a good indication of position can be deduced, except when flying very fast east or west (more than 500 kilometres [620 miles] in a day). Easily applied to the leg, these tools overcome many of the tape, glue and harness problems of the long-term deployments of PTTs and GPS loggers. The use of a saltwater switch allows recording of the times when the bird is on the water or in the air. Amazingly, on those long and fast migrations the royals were sitting on the water for about 60 per cent of the time — 'sleeping' is a necessity even for albatrosses! — and the longest continuous flight was 14 hours, with the average being about three to four hours.

The technological challenge for the future is to find a way to combine all the useful features of the various pieces of technology used so far into one package so that we can really tell, and not just assume, what the birds are doing. Is that bird recorded near a fishing fleet actually at risk, or is it just passing by, en route to somewhere else?

We have only just slightly opened the door on the life of the royal albatross during the 80 per cent of its life spent at sea. Forward-thinking researchers like David Nicholls are already working on the design of new technologies, assisted by the lessons learnt from a 'New Zealand bird that was (for a while) too fast for a French satellite'!

BELOW LEFT A pair of northern royal albatross returning to nest on The Forty-Fours in the Chatham Islands may well have circumnavigated the earth since their last visit.
BELOW RIGHT The site of much of the author's work, a dense colony of northern royal albatrosses crowds the small plateau on Middle Sister Island, north of the Chatham Islands.

Science and Conservation • 165

18
Buller's Albatross
Paul Sagar

Dr Paul Sagar specialises in long-term studies of the effects of fisheries and climate change on seabird populations for NIWA (National Institute of Water & Atmospheric Research), New Zealand. His albatross research began in 1976 with a University of Canterbury expedition to The Snares, and continues with numerous stints on New Zealand's subantarctic islands, plus a memorable trip to Kerguelen.
p.sagar@niwa.co.nz

Synopsis: A long-term study of a New Zealand endemic casts light on many details of its complex life history and breeding system, yet with each discovery more questions arise.

The chicks on nests among the *Hebe* were always the ones that I dreaded approaching. To reach each nest meant slowly crawling in the mud under the tough, sharp, twisted branches of the shrubs, and this gave the chicks plenty of time to gauge the distance and angle before initiating their best means of defence — copious projectile vomiting directed accurately at any perceived threat. The grey slurry of fetid fish and stinking squid almost invoked a similar response from me. Despite being dressed in PVC clothing, I knew that the oil in the slurry would defy gravity and seep under my wrist bands and onto my overalls. The distinctive smell would never wash out. So, why am I doing this? Why not just count the birds from a safe distance and leave them be? Well, these chicks were included in a study colony that was part of our long-term project to determine the population dynamics of Buller's albatrosses, and so they needed to be measured and banded if we were to have complete information about their survival and fidelity to the colony in years to come.

Banding the chicks allows them to be identified as individuals for as long as they live, and that could be many years. We started our project in 1992 and visited our study colonies on the uninhabited Snares Islands, south of New Zealand, each year until 2006, but the earliest banded birds did not begin returning until 1997. Subsequently, we have recaptured many of these known-age birds on each visit to the island. From this we have learned that most returned to the island when aged between six and eight years old, though we know nothing about where they spent the intervening years. However, as albatrosses have to come ashore only to breed, we presume that all those years between fledging and recapture were spent travelling the ocean learning where and when food was most abundant.

Most Buller's albatross breed for the first time when 10–13 years old, with males doing so at 11 years and females at 10.4 years, on average. Interestingly, most males return to breed at the same colony where they were born, whereas only about 50 per cent of the females do so, the remainder moving to colonies nearby. This is likely to reduce inbreeding, and so maintain genetic diversity among the population. However, once the birds breed they are extremely faithful to that colony for the rest of their lives.

When the six-, seven- and eight-year-olds first return to The Snares, they do so late in the breeding season — about the time that the established birds are hatching their chicks. Generally, these young birds stay around the periphery of the breeding colonies, interacting with others of similar status in groups (gams) of up to 20 birds. In subsequent breeding seasons these birds stay ashore for longer periods as they gain social experience. By their third year of coming ashore, the birds arrive from the end of the egg-laying period and are usually associated with a prospective nest site. Males always spend a greater percentage of the days ashore than do females. The latter appear to fly along cliff faces near the breeding colonies, occasionally landing briefly and interacting with a male with an empty nest. We found that the eggs of females breeding for the first time were narrower than those of more experienced birds, providing a useful means of identifying such birds that had not been banded as chicks.

Tracking of these known-age birds by satellite telemetry has shown a progressive development of their time ashore and a systematic familiarisation of foraging areas at sea. The sequence of behaviour that

ABOVE RIGHT Slightly smaller and darker than the author's study birds on The Snares, a pair of northern Buller's greets at their nest on the Forty-Fours off the Chatham Islands.
RIGHT Left alone on the nest while its parents are at sea, young Buller's albatross chicks are adept at warding off potential predators such as skuas (and researchers) with a squirt of pungent, well-aimed vomit.
FAR RIGHT Using vegetation and mud as mortar, an incubating bird spends many hours strengthening its pedestal nest.

develops in the two to three years that most birds spend selecting a mate and obtaining a nest site appears to involve a single brief visit ashore during one breeding season, followed by multiple brief visits interspersed by long absences when the birds forage in Tasmanian waters. The birds then lengthen the number of consecutive days ashore between long trips away, before shortening the duration of long trips, and finally lengthening the duration and sequences of consecutive short trips. During the latter stage the birds are ashore for several days, typically going to feed at distances up to 100 kilometres (62 miles) offshore each night. Being conspicuous and diligent in occupying a nest site appears to be all-important when trying to attract a mate.

An intriguing result of our research into the lives of Buller's albatrosses banded as chicks is that we have recaptured many more males than females. However, we do not know whether this is a result of different behaviours, or male bias in the proportion of chicks that fledge. We do know that before their first breeding attempt, males tend to stay ashore for longer and have a stronger tendency to nest in the same colony where they were raised. Both these behaviours would favour the recapture of more males than females. Another contributor to the male bias could be that more male chicks are produced and survive to fledge. As with most albatross species, male Buller's albatrosses are larger than females. Consequently, parents have to provide more food to raise a son than a daughter. This has led to the suggestion that both relatively inexperienced and old parents tend to raise daughters rather than sons, but the majority of the breeding population raise more sons than daughters. Another factor affecting the sex ratio of chicks would be the availability of food at sea, with years of abundance favouring males. In the past three years we have used measurements to estimate the sex ratio of the chicks banded in our study colonies, and in each year more males than females have been produced. However, we need information from several more years before we will be able to shed any light on the dynamics of sex ratios and what may change the balance.

Our banding shows that about 31 per cent of chicks survive to return to the island and once they start breeding their annual survival rate is extremely high at 97 per cent. This means that these birds can expect to breed for about 25 years, and so reach an age of about 36 years. However, some will live much longer — the oldest in our study was a female, banded as a breeding bird (and so at least 10 years old) in 1948, by Dr Lance Richdale, and last recorded, breeding again, in 1993 when at least 56 years old! So, it is rather a humbling experience to know that some of the albatross chicks that I banded in 2004 should still be flying and breeding long after I am pushing up daisies! It also highlights the need for another generation of seabird researchers to continue the project.

ABOVE The author, here weighing a known bird at his study site, has been researching the population dynamics of the Buller's albatross on The Snares since 1992.
LEFT Unlike many other albatross species, the agile Buller's on The Snares nest deep in the forest of the endemic tree daisy, *Olearia lyallii*.

19
Chatham Albatross
Paul Scofield

Dr Paul Scofield is the Curator of Vertebrate Zoology at Canterbury Museum in Christchurch, New Zealand. He has worked on seabirds from the Arctic to the Antarctic and on the Chatham Islands since 1986, as well as more recently in Chile and Hispaniola. He is co-author of the recently published Field Guide to the Albatrosses, Petrels and Shearwaters of the World *(Christopher Helm and Princeton University Press 2007).*

Synopsis: A detailed study of a critically endangered species under utterly challenging field conditions reveals many aspects of the life history, migration patterns, fisheries threats and the impact of global warming.

Situated at the very south of the Chatham Islands, facing the emptiness of the Southern Ocean with nothing between it and the Antarctic continent, lies one of the most evocatively named of all New Zealand's offshore islands, The Pyramid. Its precipitous, virtually inaccessible 1.7 hectares (4.2 acres) are the only home of the Chatham albatross, the rarest of New Zealand's 11 breeding albatross species. Only from the south is the island truly pyramidal, from other angles some say it resembles a rhino's horn. Its most prominent features are a small shelf 30 metres (100 feet) above sea level on the south-east and a shallow cave with a huge mouth on the western face. It has proven to be one of the most arduous locations to attempt long-term seabird research.

The Chatham Islands are a remote group 700 kilometres (435 miles) east of New Zealand, home to a resilient breed of folks that can take all that nature throws at them. Originally settled by the Moriori people about 600 years ago, and more recently by Maori and Europeans, nearly all make a living from fishing and farming. Consequently, the main islands have suffered greatly from the depredations of introduced predators and from habitat destruction, but the smaller uninhabited ones remain refuges for seabirds, providing crucial breeding sites for a number of rare species. Three species of albatross are essentially endemic to the archipelago, the Chatham which breeds nowhere else, and two others with more than 99 per cent of their breeding populations based there: the northern royal and the northern Buller's, possibly a new species also known as the Pacific albatross. All three species nest on privately owned offshore islets.

The New Zealand Department of Conservation (DoC) currently ranks the Chatham albatross as category B, the nation's second highest priority category for conservation management. The species is also rated internationally as Critically Endangered by the IUCN, primarily due to low population numbers restricted to only one tiny and precarious breeding site. In 1999 there were 5333 occupied nests on The Pyramid, suggesting a probable breeding population of 10,000–11,000 individuals, although previous estimates, based on counts from aerial photographs, had indicated a population of 3200–4200 breeding pairs. It is currently estimated that 1200–1500 chicks fledge in each season.

The Chatham albatross is a medium-sized species most closely related to the Salvin's and white-capped albatrosses. While feeding primarily on the Chatham Rise during the summer breeding season, they migrate to Chilean and Peruvian waters during the non-breeding season and are thus susceptible to El Niño climatic events which reduce their foraging success. Birds return from South America in the early southern spring and lay in October; hatching peaks in late November, and the last chicks fledge in May. Indications from banding studies are that, although survival is generally high (c. 92 per cent), during the last significant El Niño in 1987 adult mortality increased by 5–10 per cent. Meanwhile, in New Zealand waters, it has been found that Chatham albatrosses fall victim to low but persistent levels of bycatch, primarily in the Chatham Rise longline fisheries. It is for this reason that research originally instigated by Dr Christopher Robertson of Wild Press, Wellington, and now conducted with myself as his co-worker, has been funded by DoC, WWF and recently by the New Zealand Ministry of Fisheries.

ABOVE RIGHT A basaltic monolith with a gaping cave in its flank, The Pyramid is the Chatham albatross's only nesting site.
RIGHT Sharing a narrow stone cleft where they have started building a nest, a courting pair rests close together.

Fieldwork first began on The Pyramid in the 1970s, but more intense annual visits have been undertaken since 1998. Since then, each season we have monitored three cohorts of chicks banded in the early 1990s and a proportion of nests to establish adult survival and fecundity. Part of our research has also been to attach satellite transmitters, GPS loggers and geo-location loggers to birds to establish where they go during the non-breeding season (May–October). Six adult birds feeding chicks have been satellite tracked using Argos satellite transmitters, 10 have had geo-location devices attached and two have been GPS logged. Different birds are targeted each year to avoid undue stress to particular pairs. This research has shown that the birds migrate to the eastern coast of central South America where anecdotal accounts suggest that bycatch in the artisanal longline fisheries of Peru and Chile may be significantly impacting this vulnerable species, though the level of mortality remains unknown.

There are indications that in all Chatham Islands albatross populations the possible effects of climate change may be making their mark. Average summer temperatures in the last 10 years are the highest on record and the effects of heat stress are clearly obvious on chicks during our visits, with high levels of mortality, either from direct collapse or from individuals' inability to ward off tick-borne disease prevalent in hot weather. For example, the 2005/2006 season was a particularly bad one with only 32 per cent chick survival. Only the ongoing study of demographic parameters and special foraging regimes will lead to our understanding of how environmental fluctuations, climate change and fisheries are interacting to affect the health of this endangered species.

A field trip to The Pyramid is not undertaken lightly. Due to the constant barrage of weather fronts assailing the Chathams, it is generally only possible to land or get off the island on one or two days a month. But, when particularly bad weather patterns hit, months may go by without a calm day, and so entire field trips have had to be abandoned. The landing is treacherous across kelp-covered rocks washed by 2-metre (6.5-ft) swells even on the calmest of days — a constant reminder that this is the great Southern Ocean.

Even when finally ashore life is not simple. First, there is only one place flat enough to camp, with just two rough and bumpy, though mostly horizontal, spots for pitching tents. As there is no soil, tent pegs are useless, so pitons must be bashed into the tiny cracks in the hard andesite rock or huge fishhooks slid into the tiniest and tightest of fissures. Tent guys must have specially designed springs attached, to allow the tents to give in winds commonly reaching 60 knots. We expect at least one bad storm per trip and have an emergency plan to allow evacuation to higher ground if needed, with harnesses and ropes enabling us to physically secure ourselves to the rock to prevent being washed away in the worst of conditions.

If that wasn't enough, when it rains the smooth volcanic rock, coated in a fine layer of guano, becomes so incredibly slippery that movement on the steep slope becomes impossible. Such is the danger of getting stuck out on some precipitous flank, that the mere sight of rain in the distance, or any dampness in the wind, is enough to have us scuttling back to camp as quickly as possible.

Make no mistake, the island is steep. We joke that we have to alternate the direction we leave camp on our daily searches for banded birds lest one leg should grow a foot longer than the other to adjust to the precipitous angle! On dry days, though, with the aid of a rope, there are very few nest sites that can't be visited by trained rock climbers. While it is clear that this is not a job for the faint-hearted, nor those with vertigo, the findings have enabled us to construct a realistic picture of this rare bird's life cycle.

ABOVE High on The Pyramid, protruding from the cold sea fog below, Chatham albatrosses court on a sun bathed ledge.

LEFT With the smallest breeding area of any albatross species, the Chatham albatross depends on scant vegetation growing from fissures on the steep face of The Pyramid as a base to attach their nests.

20
The Plight of the Albatross on Tristan da Cunha
Conrad J. Glass

The author's book *Rockhopper Copper* (The Orphans Press, Leominster, England, ISBN 1-903360-10-2) tells in more detail of his work to conserve wildlife and the environment in and around Tristan as well as recounting everyday life and something of Tristan's history. It is available by post for £13 (or equivalent, by undated cheque) from: C. Glass, Thompson Street, Edinburgh of the Seven Seas, Tristan da Cunha, South Atlantic Ocean TDCU 1ZZ (e-mail conrad.glass@gmail.com) or the publishers, info@orphanspress.co.uk.

Conrad J. Glass is Police Inspector and Conservation Officer on Tristan da Cunha, the world's remotest inhabited island. He is a direct descendant of the original settlers who founded the small fishing community nearly 200 years ago, now numbering 280 residents.

Synopsis: An inside view of the emerging conservation consciousness of an isolated island community where three species of albatrosses nest, recounted first-hand by a native-born Tristan islander.

High above the white Atlantic clouds, a very long time ago, an albatross soared with effortless grace, lazily lifting and falling as it rode the updraughts from warm air currents, its movements similar to those of a boat out on the sea. This was a magnificent bird with a 2-metre (6.5-ft) wingspan. The top of its wings were black from tip to tip, as was the end of its tail, the rest of its body white, with a hint of grey at the back of its neck and head. The long slim aquiline beak, black on the sides with a yellow strip on top, ended with a pinkish hue on the curved tip — an amazing contrast of colours to find on an albatross. It was this that ultimately gave this special variety its name: the yellow-nosed albatross.

The albatross was returning from foraging, having travelled hundreds of miles out to sea. Its destination: a volcanic archipelago, thrusting up in the middle of the South Atlantic Ocean. The largest island, towering 2060 metres (6760 ft) above sea level, with lofty cliffs rising sheer from the sea, was often hidden in swirling clouds and mists. The slopes were covered by lush foliage of long tussock grass, ferns, bushy trees, moss, lichen and a variety of other plants. Among this green tangle, the albatross built its nests and reared its young in peace, sheltered from the buffeting winds. Later the island was to be called Tristan da Cunha by other wanderers coming across the oceans, the European explorers. But until their arrival, there was nothing here to harm albatrosses or their fledglings.

As the albatross neared Tristan, it saw some objects close inshore. Instinctively it dived, covering the distance in seconds. The objects were sailing ships. It had seen ships before, but not this close to Tristan. The albatross flew within feet of the masts, watching as men scrambled about the rigging. Losing interest, it veered away, landing on a plateau high on the mountain. The year was 1506. The threat to its way of life had arrived. These were humans: the mammals that changed the world.

This was the year the Portuguese explorer Tristão da Cunha discovered the island that now bears his name. For the next two centuries visiting pirates, whalers, sealers, merchant and naval ships hailing from America and Europe all left their mark on Tristan. Some captains put goats and pigs ashore, which soon multiplied, destroying albatrosses and vegetation. Wild pigs are not fussy; albatrosses were as much part of their menu as grass roots. Hungry sailors, scavenging pigs and herds of goats together ended the albatross colonies that once covered the lush plains near the shore. Surviving birds built nests higher up on the mountain.

August 1816 was when the first families settled on Tristan, including my great-great-grandfather seven generations removed. They cleared land for farming, brought in cattle and poultry, and hunted all the wild pigs and goats. They also took albatrosses as a source of food. Over the years albatross numbers began to diminish. The Tristanians had to travel further afield to reach the remaining birds, so they soon realised that they had to do something to sustain the decreasing number. A policy was endorsed that if albatrosses were killed for food, it was by a quota system for specific reasons. A man who was building or roofing a house, or enclosing stone-walled allotments, was allowed to kill 25 or 30 yellow-nosed albatrosses to feed the

BELOW Historic painting by Augustus Earle: 'A man killing an albatross — Summit of Trista De Acunha', around 1824.

people helping him. This was done as a means of payment because there was no money on Tristan (and indeed there was none until some 60 or so years ago). Today the custom continues, but using lamb or beef. Yellow-nosed albatross were also consumed on the Easter Weekend as 'special roast'. Two birds only per person were allowed to be slaughtered. This policy was adhered to by those who looked after the affairs of the island: the Anglican Padres, Chief Islanders and later by the Administrators. Very occasionally, permission to slay a few yellow-nosed albatrosses was granted to people on outings away from the settlement, but this was only allowed with the consent of the Chief Islander or his nominee.

The sooty albatross was hardly ever hunted because it nested on inaccessible cliff faces on the mountain. However, the Tristan albatross, a type of Wandering albatross called gonia by the islanders, was just about decimated by human exploitation. Due to its size, fewer were needed to satisfy one's needs but this gave it top rank on the food list. With a slow reproduction rate, rearing but one chick every two years, the gonia were killed off more rapidly than they could breed.

The year 1882 saw another threat to the albatrosses: rats came to Tristan through another human folly. It happened when the sailing ship *Henry B Paul* was deliberately run ashore to claim insurance. Rats became a menace to birds and people alike — and were here to stay.

The last time gonias are known to have nested on Tristan was in 1950, high on the mountain, at a place called Round Hill on the south-western side. Two birds were observed, but disappeared amid speculation that a couple of islanders had killed them. A law was passed to protect them, but this was too late. More recently several have been seen flying about the southern part of Tristan, fostering growing hopes that gonias may have returned to breed unseen on the rugged and nearly impenetrable far side of Tristan.

In 1976 the Administrator and the Island Council passed a law stating that the yellow-nosed albatross was now a protected bird. With each family allowed a sheep and cattle quota, it was deemed the islanders no longer needed to kill this bird for food. The sooty albatross had to wait for legal protection until 1986 when, by decree of the Island Council, all albatrosses in the archipelago became fully protected.

In 1993 the Tristan Government's Natural Resources Department was created, with James Glass at its head, responsible for conservation matters on and around the islands. Today a conservation section has been added, with the appointment of the first full-time Conservation Officer, Simon Glass, and the Tristan authorities are making plans to deal with the rats. By far the biggest threat to albatross remains from human activities, especially longline fishing and from bits of plastic debris ingested by birds mistaking them for food. Dead albatrosses found floating in the waters around Tristan have suffered the effects of human activity elsewhere.

So do Tristanians still kill albatrosses? I like to think not. There may be opportunistic poaching of a couple of chicks from time to time. In this isolated but self-reliant community of sturdy independent-minded people such things are possible, carried out far from those of us determined to uphold the conservation laws and protect all wild creatures whose life, like ours, depends on Tristan da Cunha's unique environment.

ABOVE LEFT Atlantic yellow-nosed albatrosses, now strictly protected, still nest within sight of the village that once depended heavily on albatrosses for food; the larger Tristan albatross vanished long ago.
ABOVE RIGHT Clinging to the foot of the Tristan volcano, Edinburgh of the Seven Seas — population 280 — was founded in 1816 and is the most remote island community in the world.
BELOW Tristan da Cunha Island (with a giant petrel flying by) is a dormant volcano straddling the Mid-Atlantic Ridge and last erupted in 1961. Both sooty and yellow-nosed albatrosses nest on its slopes, while the 2060-metre (6758-ft) summit rarely emerges from the mists.

Science and Conservation • 171

21
Southern Seabird Solutions Trust: Conservation through Cooperation

Janice Molloy, Convenor, Southern Seabird Solutions Trust
John Bennett, longline skipper, Sanford Ltd
Caren Schroder, Marine Manager, WWF-New Zealand

Janice Molloy is the founder and convenor of the Southern Seabird Solutions Trust, working closely with Captain John Bennett, an active longline skipper with the New Zealand fishing company Sanford Ltd and Caren Schroder, Marine Manager of WWF-New Zealand. Their joint efforts are driven by their conviction that albatrosses can be spared through innovative collaboration. www.southernseabirds.org

ABOVE The Buller's albatross, a bold scavenger around fishing vessels, is one of the beneficiaries of the Southern Seabirds Solutions work to educate and inspire fishermen in adopting seabird-friendly techniques.
RIGHT Working under the cover of darkness, a longliner sets his gear at night, when albatrosses cannot see the tempting baited hooks before they sink out of reach.

Synopsis: An innovative group brings together fishermen, conservationists and government officials to find solutions to seabird bycatch problems, and to spread the word that, with enough effort, successes can be achieved.

Many albatross populations are in trouble, and one key threat they face is from fishermen accidentally catching them on fishing lines or on the wire ropes of trawl nets. Fishermen have been viewed in two different ways over this issue in recent years. One view has been to assume fishermen know the damage they are causing, and are not concerned enough to do anything about it. This has resulted in governments controlling fishermen's activities through regulations, and environmental groups campaigning against fishing practices. Another view has been to regard fishermen as not fully aware of the impact of their activities, and to get alongside them and educate them. This approach assumes that if these people have a better understanding, they will want to change their fishing practices and champion 'seabird smart' fishing.

Both approaches have been used in New Zealand. In the early 1990s, when fisheries impact was first becoming known, there was little or no constructive dialogue between the fishing industry, government and environmental groups. Most efforts were put into discrediting the other 'side' rather than addressing the problem (today some environmentalists still remain opposed to engaging with even the proactive elements of the industry). However, this period was important, because it raised the issue in the public eye and made the fishing industry more accountable. Around this time, some fishermen, recognising the detrimental effects on albatross populations, began experimenting with different fishing methods, such as dyed baits, weighted lines and setting baited hooks under water. By stepping forward and seeking advice and support, these individuals brought about a fundamental change in relationships because all fishermen were no longer lumped together as a faceless group that could be written off as uncaring scoundrels. The more collaborative government officials and environmental groups, such as WWF, showed willingness to work with these fishermen. This also meant fishermen themselves could no longer hold the view that these organisations were using the seabird issue to close the fishing industry down. It was a turning point for many people.

In 2002 mutual trust had reached a level enabling the various parties to come together in one room and agree to solve the accidental deaths of seabirds in a cooperative manner. It was at this meeting that the group Southern Seabird Solutions (SSS) — which has since become a charitable trust — was formed, made up of government agencies, environmental NGOs, seabird researchers, tourism operators and, above all, fishermen and fishing companies. The trust is based in New Zealand but has a southern hemisphere-wide scope, because of the migratory nature of albatrosses and the need to solve the problems throughout their range. The three authors are longstanding members of the trust.

While there are not infrequent tensions between these various parties in other areas of fisheries management, we all 'leave our swords at the door' when we come together for SSS meetings. Engendering this trustful and cooperative approach between our partners has been the cornerstone of the group's success.

Most SSS projects involve fishermen, simply because receptiveness is greatest among peers. Fishermen can also more easily stand in their colleagues' shoes, knowing how to communicate the message in a practical and meaningful way. For instance, the group

has to date carried out three skipper exchange projects, with Chile, Peru and Reunion Island, all with excellent outcomes. In the Chile exchange, the skipper returned home after a month aboard John's longliner, motivated to continue spreading good practice in his own fleets. He has attended workshops around Chile talking about his experiences in New Zealand and describing the measures he observed being used. In the Reunion Island exchange, the whole fleet has since begun using a new weighted longline that sinks quickly beyond the diving range of seabirds. In the Peru example, the New Zealand skipper was able to suggest some new solutions that could have potential with the type of fishing methods used there.

Both fishermen and companies need to feel good about themselves, and through their adoption of 'seabird-smart' fishing techniques strive to act as 'good citizens'. So with the combined assistance of environmental NGOs, industry and government agencies, SSS has promoted industry-wide mitigation efforts. Successes are celebrated publicly through the general news media, and via seafood trade publications. Conscientious fishermen now use a range of 'seabird smart' fishing techniques, depending on their fishery. Examples include reducing the attractiveness of trawl and longline vessels by the reduction of offal and bait discarded into the water, preventing seabirds from flying close to the stern of vessels by using bird streamer lines and other physical barriers, and lessening the risk of seabirds encountering a hook by setting at night or weighting longlines.

During its first three years, SSS focused much of its efforts on the communication of successes, and most particularly conveying these within industry circles. Monthly stories were carried in *Seafood New Zealand* magazine, celebrating role-model skippers, new 'seabird-smart' fishing technologies and information on the birds themselves. The articles had a huge effect in building and maintaining support from across the fishing community and wider industry.

A classic example was the positive reception a new seabird mitigation advisory officer (a fisherman himself) received when visiting a fleet for the first time. All the skippers welcomed him aboard their vessels, eager for help in improving their fishing methods, having read about him and the 'seabird-smart' concept in *Seafood New Zealand*.

SSS recently ran a number of workshops for inshore longliners at ports in northern New Zealand, hosted by their local fish-receiving shed. We brought several role-model fishermen from other fleets as speakers, as well as an environmental representative knowledgeable in seabirds and their conservation. Part of our purpose was to thank fishermen for their efforts to date, and especially to encourage them to continue using 'seabird-smart' practices. The workshops were all held in the local bars, which ensured good attendance, and participants received T-shirts with our logo and catch phrase 'Conservation through Cooperation', all creating an upbeat atmosphere of receptiveness.

Similar workshops are targeting trawler-specific issues, with multinational assemblies of fishermen, fishing industry representatives, scientists, government officials and environmentalists together brainstorming long-term practical solutions to reducing global seabird bycatch in trawl fleets.

Other fisherman-to-fisherman work has included the production of a 'seabird-smart' fishing video, hosted and narrated by a fisherman, and largely featuring skippers and vessel managers from different fleets talking about the issue. The video proved hugely popular, with copies distributed free among New Zealand fleets. A Spanish translation features a special introduction from the Chilean fisherman involved in the skipper exchange programme.

One of our messages is that even occasional catches of albatrosses will have an impact. A fisherman who catches one or two albatrosses a year may not appreciate the seriousness of the event, but when he understands that literally thousands of fishing boats throughout the southern hemisphere are doing likewise, and sometimes killing many more, the consequences become clear.

A critical factor in the success of the Southern Seabirds Solutions Trust has been developing a goal that everyone can agree on and work towards. In addition, gathering this initially loose coalition of interest groups under a name and governance structure has resulted in a cohesion and identity that members are proud to be part of. The main limiting factor to date has been securing enough resources to undertake the many additional projects we have lined up.

BELOW Simple 'tori', or bird lines, dangling above the taut cable of a deep-sea trawl operation are sufficient to keep huge flocks of black-browed albatrosses out of harm's way as they vie for scraps in the wake of a fishing vessel working the South Atlantic.

22
The Albatross Task Force: A Sea Change for Seabirds
Ben Sullivan

BELOW The simple technique known as 'tori lines' ('tori' meaning 'bird' in Japanese) is used to scare birds during trawling or line setting by longliners. Streamers hung at 5-metre (16 ft) intervals from a 50-metre (164-ft) line with a float dragging at the end is all it takes, although wind and high seas can reduce its effectiveness. *Diagram adapted from BirdLife International.*

Dr Ben Sullivan is based in Tasmania and works for the Royal Society for the Protection of Birds (UK), involving policy development and coordination of the BirdLife Global Seabird Programme, and management of the Albatross Task Force. A former at-sea fisheries observer with Falkland Conservation's Seabirds at Sea Programme, he has also worked on aspects of implementation of the FAO International Plan of Actions-Seabirds, and remains passionate about bycatch mitigation techniques for both longliners and trawlers. Ben.Sullivan@rspb.org.uk

Synopsis: A hands-on report of the successes of BirdLife International in implementing seabird bycatch mitigation techniques among fishermen in the most affected regions of the world.

There was a resounding shudder as the winches kicked in, followed by a piercing alarm to wake the great unwashed and over-worked. It was 3 a.m. and as I tried to slide out of my bunk in a sleepless haze the vessel was hit by a Southern Ocean roller that threw me across the cabin and slammed me against the bulkhead. After screwing my head back on and circumnavigating the cabin a few times trying to get my legs into my freezer suit, I stepped out on deck to have my breath taken away by an icy southerly buster. The deck crew were huddled-up like emperor penguins ... waiting for orders that they hoped would never come. With the deck lighting and horizontal sleet creating almost whiteout conditions, I could just make out the shadow of a massive bird banking around the stern of the vessel ... this is the theatre of the albatross.
— notes from my Fisheries Observer diary.

Albatrosses and fisheries are inextricably linked throughout the world, and have been since humans first took to the sea. Like albatrosses, which conduct some of the most amazing migrations on earth, massive migratory fishing fleets also ply the world's oceans. These compete in space and time with albatrosses for resources, luring many into potentially fatal interactions, to the point where three decades of modern longline and trawl fishing is threatening millions of years of evolution.

The BirdLife Global Seabird Programme was established in 1997 to address a broad range of issues but its main coordinated focus to date is exemplified by BirdLife's 'Save the Albatross' Campaign, which addresses mortality caused by longline and, more recently, trawl fisheries. In broad terms, the programme focuses on local, regional and international advocacy to raise awareness of these concerns within the fishing industry and wider community, and to facilitate implementation of onboard mitigation measures to reduce the level of seabird mortality.

In mid-2004, the BirdLife Global Seabird Programme recognised that in addition to national and international efforts to halt the rapid decline of

RIGHT When fish-processing offal is being discarded from a South Atlantic trawler, enormous numbers of albatrosses and petrels are attracted to the vessel's wake, coming dangerously close to the trawl cable during fishing operations working without deploying 'tori' lines.

174 • Science and Conservation

albatross populations, the most urgent gap was resources and capacity to deliver international 'grass roots' action to demonstrate to fishermen the range of simple and cost-effective mitigation measures available to reduce albatross mortality to negligible levels. Many programmes around the world place observers on vessels to monitor and record seabird bycatch. However, there is a real lack of people with practical know-how to work with fishermen to demonstrate appropriate 'best-practice' mitigation measures, and who have the personal skills and dedication to live and work at sea. Hence, the Albatross Task Force was born.

BirdLife and the Royal Society for the Protection of Birds (UK BirdLife partner) launched the world's first international team of mitigation instructors working with fishermen both on shore and at sea. Their task was to ensure the adoption of an effective suite of mitigation measures onboard longliners and trawlers. Instructors are placed in teams around the Southern Ocean to target bycatch 'hotspots' where albatrosses are most at risk. The placement of teams is based on a range of criteria, including: (1) regions where bycatch tallies clearly demonstrate that a problem exists, (2) regions where the BirdLife Global Procellariform Tracking Database (see *Tracking Ocean Wanderers*, BLI 2004) highlights a significant spatial and temporal overlap between albatross distribution and fishing effort, and (3) developing countries which experience difficulties in resourcing appropriate seabird bycatch initiatives.

The Task Force was launched in 2006 in Cape Town, with two full-time instructors working in pelagic longline fisheries and one in trawl fisheries. Since then, we have expanded to have teams in Argentina, Brazil, Chile, Namibia and Uruguay, and we have plans to expand further in South America and southern Africa, and beyond!

In many fisheries we have the tools to reduce seabird bycatch to almost zero, particularly in demersal (bottom) longline fisheries. In this industry the work of the Commission for the Conservation of Antarctic Marine Living Resources (CCAMLR), the body responsible for the management of Antarctic and subantarctic fisheries, has demonstrated that with political will and technical knowledge seabird bycatch can be eliminated. While we are not as advanced with mitigation measures for pelagic (open sea) longline fisheries, there are several very promising research programmes under way to refine and develop new measures for these pelagic fleets. There is no reason why the same result cannot be achieved for pelagic fisheries' vessels the world over, whether operating in national waters or on the high seas. The same can be said for trawl fisheries.

The aim of the Albatross Task Force is to reduce seabird bycatch in fisheries around the world. Currently we are focused on issues in national waters, but we recognise the need to work with the large, distant-waters fleets that operate on the high seas (which constitute two-thirds of the world's oceans). These massive fleets, which primarily represent pelagic longline fisheries targeting tuna and swordfish, are critical to albatross conservation. BirdLife is working closely with the international bodies responsible for the management of these fleets (Regional Fisheries Management Organisations) to assist them in developing similar observer programmes that address seabird bycatch.

A range of political, technical and management measures are required to turn international and national policy and regulations into 'grass-roots' action that will halt the rapid decline of many albatross populations. The Albatross Task Force delivers what is most urgently needed: action where it matters most … at the stern of the vessel.

BELOW LEFT A mêlée of black-browed albatrosses and giant petrels follow in the wake of a longliner working the South Atlantic. Unless mitigation techniques are applied, such as bird-scaring streamers or night setting, many are hooked and drowned as the lines are deployed.

BELOW RIGHT In a badly timed landing, this white-capped albatross passing underneath the taut trawl warp cable will almost certainly be sliced and dragged down as the ship heaves in heavy seas.

23
South American Perspective: Fisheries Mortality
Marco Favero

Dr Marco Favero is an Argentinian seabird ecologist, Head Researcher at the National University in Mar de Plata and the National Research Council for Science and Technology (CONICET). After 20 years' research in Antarctica and the Southern Ocean, he now serves with CCAMLR on seabird bycatch issues, on the board of ASOC and the ACAP Advisory Committee. mafavero@mdp.edu.ar

BELOW At opposite ends of the world, albatrosses on their nesting islands carry evidence of close calls with longline operations: a critically endangered female Tristan albatross incubating on Gough (left) and a southern royal hatching its chick on Campbell (right). Both species come into close contact with the pelagic fishing fleets working in South American waters.

Synopsis: A technical survey of the state of albatross mortality in South American fisheries indicates considerable concern, especially in areas influenced by the Subtropical Convergence in the south-west Atlantic or the Humboldt Current in the South Pacific, where frontal systems concentrate both fishing effort and albatross activity.

The South American continental shelf, covering about two million square kilometres (780,000 sq miles), is an area of exceptionally high primary productivity due to the influence of very rich subantarctic waters and the dynamics of diverse frontal systems (Acha et al. 2004). As a result, food is very abundant for large numbers of top predators visiting these waters at different stages of their life cycles (Croxall and Wood 2002). Also, as a consequence of these concentrated living resources, the region is extensively exploited by different fishing fleets on both the Atlantic and Pacific sides.

Although other areas of the southern hemisphere may be more significant in terms of albatross diversity, the waters off southern South America are probably the world's most important with regard to biomass, especially considering that a major proportion of the most numerous of all albatross species, the endangered black-browed albatross, breed in islands of the south-west Atlantic and south-east Pacific Oceans. Wandering albatrosses and grey-headed albatrosses also breed here in lower numbers, while the area is visited by many other species coming from distant islands. For example, waters off southern Brazil, Uruguay and northern Argentina are frequented by Tristan albatrosses breeding on Gough Island. Many other species that breed around New Zealand are regular visitors off Peru and Chile, and in some cases also the south-west Atlantic, such as Salvin's, Buller's, southern and northern royal, and the critically endangered Chatham albatross (Tickell 2000, BirdLife International 2004).

The conservation status of such abundant species as the black-browed albatross has changed over the last decade due to steep declines observed in colonies at Malvinas (Falkland) and South Georgia Islands, with a loss of around 90,000 nests estimated from the former alone (Huin 2001). These population decreases, also observed in other species such as the wandering albatross, have been at least partially attributed to both adult and juvenile mortality associated with fisheries practices (Poncet et al. 2006). Concerns over this issue were established in South America at the Workshop for the Conservation of Albatrosses and Petrels in Uruguay (BirdLife International 2001), with the participation of most countries from the region, and reinforced in the FAO/BirdLife Workshop in Chile (Lokkeborg and Thiele 2003). Nowadays, of the 11 ACAP member countries, five are from South America: Argentina, Chile, Ecuador and Peru have ratified the agreement, and Brazil is in the process of ratification (Uruguay will hopefully join in the near future). Following is a country-by-country summary of the current situation.

In Peru and Galapagos the artisanal fishery is large and increasing, with some vulnerable and critically endangered seabird species affected by association with fishing vessels (Goya and Cárdenas 2003). Between 2000 and 5000 albatrosses are believed to be killed annually by Peruvian longliners. Not only incidental but also intentional mortality has been recently reported, threatening species like the Galapagos waved albatross (Awkerman et al. 2006).

The Chilean longline fishing fleet targeting Patagonian toothfish, swordfish and hake, with a yearly effort of 25 million hooks, kills around 2000 albatrosses among other tube-nosed seabirds. Bycatch rates range between 0.030 and 0.047 birds per 1000 hooks, depending on the target fish species

(Moreno et al. 2003, 2006). No detailed information is available for trawlers but informal reports suggest the occurrence of mortalities from hake fisheries.

Longliners operating in Brazilian waters are well recognised as a danger for albatrosses. With fishing effort having increased dramatically in the last two decades, the offshore and artisanal fleets are by far the largest in South America. Mortalities between 0.09 and 0.12 birds per 1000 hooks were reported in mid-water longliners, and of 0.30 birds per 1000 hooks in demersal longliners. It is estimated that around 10,000 birds die per year, among which are black-browed, Atlantic yellow-nosed and Tristan albatrosses, all endangered species (Olmos 1997, Olmos and Neves 2003). Also in these waters, driftnets and demersal gillnets are other potential hazards to foraging albatrosses (Alvarez Pérez and Wahrlich 2005).

The alarming mortality rates reported in Uruguayan longliners of 4.7 birds per 1000 hooks (Stagi and Vaz-Ferreira 2000) should be considered as occasionally elevated. Updated and more robust information shows an average of 0.26 birds per 1000 hooks, mostly in a critical area associated with the convergence of Malvinas and Brazil currents (Jimenez and Domingo, in press). Other fleets, such as trawlers and jiggers, are mentioned as threats but mortality figures have not yet been quantified (Marin 2003).

With annual fishing efforts of roughly 30 million hooks, the longline fishing fleet operating in Argentinian waters has three primary targets: the Patagonian toothfish, the kingclip and the yellownose skate. In keeping with Brazil, Uruguay and Chile, the main bird species affected here are black-browed albatrosses and white-chinned petrels (Gómez Laich et al. 2006, Gandini and Frere 2006). The average capture rate is 0.04 birds per 1000 hooks, and the minimum annual mortality was estimated between 2100 and 4200 birds (Favero et al. 2003). On the other hand, the large and complex Argentinian trawl fishery, with its diverse fishing methods, targets, and seabird species affected, presents a challenging landscape for researchers and conservationists. In small and medium-sized trawlers black-browed albatrosses and southern royal albatrosses, as well as petrel species, interact and die in entanglement or collision with warp cables (González Zevallos and Yorio 2006, Favero et al. unpubl. data). Furthermore, preliminary information indicates that albatrosses feeding along the continental shelf-break also interact with the many squid-jigging vessels operating in those waters. Although more information is definitely needed, in view of the vast scale of this fishing effort, even a very small mortality per vessel may result in a significant impact on albatross numbers.

All available information shows strong seasonal variability and important environmental and operative factors affecting albatross mortality. Some interesting mitigation techniques were recently developed for coastal and offshore longliners and trawlers. Good examples can be found in Argentinian, Brazilian and Chilean fleets, with very promising results. Some are adaptations of methods widely used elsewhere (such as streamer lines, night setting), plus novel test measures likely to be implemented (Sullivan et al. 2006, Seco Pon et al. 2007, González-Zevallos et al. 2007).

Researchers, conservationists and governments are joining in the search for bycatch solutions and the development of responsible fisheries in most countries mentioned, leading to the signing and ratification of ACAP and the recent formulation of National Plans of Action for reducing the incidental mortality of seabirds. Regional initiatives developed by local and international NGOs provide critical support for the development of research and conservation projects, among which the implementation of the BirdLife Albatross Task Force and the creation of marine protected areas seem to be the best approaches (Preikshot and Pauly 2005). With all of these efforts, the near future will hopefully see South America having a well-developed body of regulations for responsible fisheries, providing a friendly ocean for albatrosses to fly over forever — a vision embodied in Chilean writer Pablo Neruda's famous poem:

The wind courses the high seas
steered by the albatross,
this is the craft of the albatross:
gliding, falling, dancing, climbing,
hanging motionless in the dark light,
touching the pillared waves,
he nestles in the seething mortar
of the unruly elements,
crowned by salt,
while the frantic foam hisses,
the albatross skims the waves
on his great symphonic wings
scribing upon the tempest
a tome forever soaring:
the edict of the wind.
(Free translation from the original Spanish by Tui De Roy.)

24
Conserving Magnificent Flyers: A Personal Journey with Southern Albatrosses
John Cooper

Dr John Cooper is a marine ornithologist based in Cape Town, South Africa, whose work at intergovernment levels has been instrumental in galvanising action to protect albatrosses. His other passions include oceanic islands, racing his bicycle, and his pet cats. Now retired, he is concentrating his efforts on the conservation of the Tristan da Cunha group of islands in the South Atlantic. jcooper@adu.uct.ac.za

Synopsis: A fervent advocate of albatross preservation tells the story of his life achieving major conservation successes through field research and international diplomacy.

In 1970, at the age of 23, I moved from a land-locked country to live beside the ocean. Based on Dassen Island, off the west coast of South Africa, I studied African penguins and bank cormorants for two years, but rarely glimpsed oceanic seabirds. My daily pre-breakfast routine was to walk to the tip of Boom Peninsula to check nest contents, often beachcombing the wrack for interesting objects. On one such occasion an unusual seabird corpse caught my attention. With no seabird field guide then available, I managed to identify it by bill shape and markings as that of a light-mantled albatross. This became the second recorded specimen for the African Continent, later documented in my first-ever publication (all eight lines of it!) on this group of birds.

Although my interest had been aroused, it was some years before I had the opportunity to study albatrosses at their southern homes. I first visited subantarctic Marion Island in the southern Indian Ocean in June 1979 with the South African National Antarctic Programme (SANAP). On this and subsequent visits in the 1980s, together with my colleagues, I set up three long-term study colonies of wandering albatrosses, using the experience of my Dassen Island research as a guide. With the wandering albatross being a biennial breeder and approximately 200–250 nests being marked each year, the study has followed the individual fortunes of over a thousand birds. It was (and still is) a delight to make the round of the study nests, mark them with numbered stakes and record the adults' colour-bands. To sit down nearby until a wanderer notices you and walks across to inspect this strange newcomer from a metre (3 ft) or less distance remains for me a thrilling experience. Like Robert Cushman Murphy, the famous seabird researcher and writer of the 1920s, 'I now belong to a higher cult of mortals, for I have seen the albatross!' Every time I visit these study birds I gain pleasure in the knowledge that many of them are still alive from those early years, and that the chicks being banded now have a sporting chance to outlive me as I enter my seventh decade. Indeed, some of the banded study birds are reckoned to be a half-century or more old. These colonies, now with birds mostly of known age and breeding history, have also been used for several other research purposes, such as for dietary, foraging and human-disturbance studies.

But all was, and still is, not well with the wanderers of the Southern Ocean. The mortality caused by drowning on longline hooks is now well known. In late 1997 I was asked by BirdLife International to set up and run an international project to address this problem. Thus, the Seabird Conservation Programme was born, with the two main aims of raising awareness of the harmful effects of longlining and proposing ways of mitigating such effects. In frequent communications with national BirdLife partners around the world, an international task force was convened, with researchers and advocates employed in several countries. Our campaign was launched at the British Birdwatching Fair in 2000: 'Save the Albatross: Keep the World's Seabirds off the Hook'. Our contribution towards the United Nation's Food and

RIGHT Shedding the last of the down that kept it warm during the long winter months, a young Tristan albatross prepares to fledge from the high, windy moorlands of Gough Island in the South Atlantic.

Agriculture Organisation (FAO) International Plan of Action for Reducing Incidental Catch of Seabirds in Longline Fisheries was a notable early success of this campaign. Positive interaction between the FAO, an inter-governmental organisation, and BirdLife International, a non-governmental organisation, has set a way of working with the relevant management authorities and organisations, and with — rather than against — the fishing community. These were exciting times for me, upon which I look back with great pleasure and with some pride.

In 1999 I also became involved with the conservation of albatrosses and their close relatives at inter-governmental levels. With university colleagues, I persuaded South Africa to nominate the two species of giant petrels and the five large *Procellaria* petrels to the Bonn Convention on Migratory Species. With these species added, the initiative led by Australia to develop an international agreement for the conservation of albatrosses was expanded, eventually spawning the Agreement on the Conservation of Albatrosses and Petrels (ACAP) in February 2004 — with South Africa as a Founding Party. ACAP aims (as set out in its Action Plan: www.acap.aq) to work towards a reduction in fishery-induced mortality, the eradication of introduced predators at breeding sites, the mitigation of the effects of human disturbance and habitat loss, and the adoption of measures to reduce marine pollution. At the time of writing, 10 Parties have ratified the Agreement, and its Advisory Committee and four working groups have commenced laying out a 'road map' for the conservation of the species it covers.

Meanwhile, I have been conducting field trips to Gough Island in the South Atlantic for many years, and in 2002 was worried to learn that the introduced house mouse was found, completely unexpectedly, to be attacking chicks of the Endangered Tristan albatross. This was causing such an alarming level of mortality that breeding success in several parts of the island was reduced to single-digit figures. Out of this concern has grown collaboration between the Government of Tristan da Cunha (in whose territory Gough falls), the Royal Society for the Protection of Birds (RSPB) in the UK and my university in Cape Town. The management team has recruited experts to assess the feasibility of eradicating mice by dropping poisoned bait from helicopters. The difficulties (and costs) are huge but I very much hope that the mice of Gough will be gone in my lifetime.

At the end of January 2007 I retired from employment with the University of Cape Town — after over 35 rewarding years of working with seabirds, particularly albatrosses. What of the future? In my role as Vice-Chair of its Advisory Committee I hope to be able to stay involved with ACAP for a few years yet. The organisation has commenced a major drive to interact with the various Regional Fishery Management Organisations (RFMOs) that manage fisheries on the high seas, where so much albatross mortality occurs. Its aim is to persuade

LEFT The result of patient nocturnal work by researchers dedicated to exposing the evidence, a digital capture from a nighttime video reveals the gory action of ordinary house mice consuming a live Tristan albatross chick on its nest.

the RFMOs and their national members to adopt mitigation measures, such as night-setting, integral line-weighting and bird-scaring lines, all techniques that have been proven to drastically reduce seabird mortality in longline fisheries. I see this initiative as the best hope for the world's albatrosses, and it will be an honour to continue to support it. Albatrosses have given me, and continue to give me, much pleasure, and I have long felt both lucky and privileged to have studied and help conserve these magnificent flyers of the Southern Ocean.

BELOW A clean skeleton by the side of its nest is all that remains of a nearly-full grown Tristan albatross chick, killed by oversized introduced mice during the winter months.

25
Applying Spatially-explicit Measures for Albatross Conservation
David Hyrenbach

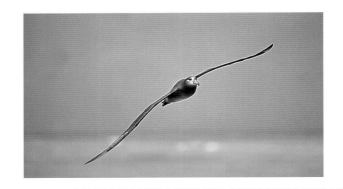

Dr David Hyrenbach obtained a PhD in oceanography from the University of California in 2001, is a professor at Hawaii Pacific University, and is affiliated with Duke University and Oikonos–Ecosystem Knowledge. He uses satellite-tracking and vessel-based surveys to study the movements and foraging grounds of far-ranging seabirds susceptible to high-seas fisheries bycatch. khyrenbach@hpu.edu

Synopsis: A technical overview promoting integrated and wide-ranging management tools, including marine protected areas and other regulations, to achieve optimum albatross protection across entire ocean basins and throughout their life cycles.

Recent technological and conceptual advances, including the advent of satellite remote sensing of ocean habitats, the design of electronic tags to track the movements of marine organisms, and the development of Geographic Information System (GIS) visualisation and analysis tools, are helping researchers to study the habits and habitats of far-ranging albatrosses. The resulting improved understanding of albatross distributions has important conservation implications, by helping to identify the locations and time periods when these species overlap with potential threats.

This enhanced ability to map the movements and habitats of protected species is at the centre of a growing 'spatially-explicit' approach to marine conservation. In this essay, I explore the potential application of this novel approach, based on the mapping and regulation of human activities in time and space, to advance albatross conservation. I advocate an integrated strategy, whereby focused and diffuse management measures are used in conjunction throughout entire ocean basins.

Marine Protected Areas (MPAs) networks are increasingly being used to manage fisheries, and to protect threatened species and marine habitats around the globe (see diagram left). Although most MPAs have focused on sessile and sedentary organisms (e.g. coral reefs, mangroves, reef fishes), there is growing interest in extending their application to the conservation of highly mobile species (e.g. marine mammals, birds, turtles). Increasingly, there are calls for the creation of large-scale oceanic reserves, akin to the parks established to protect large terrestrial vertebrates and their habitats (Norse et al. 2005).

In principle, MPAs may afford protected species with protection from some anthropogenic (man-induced) impacts, during certain periods of their life cycle. However, MPAs will not provide far-ranging albatrosses with comprehensive protection, since they routinely cover thousands of kilometres of open ocean in search of food for their chicks, and engage in vast post-breeding migrations. Thus, the implementation and enforcement of large-scale reserves capable of encompassing the entire marine ranges of albatrosses is logistically and politically unattainable. Nevertheless, MPAs may prove feasible during certain critical periods of the albatross life cycle, particularly in those instances when these species aggregate at specific habitats defined by bathymetric (e.g. continental shelf-breaks and slopes) and hydrographic (e.g. frontal systems) features to forage (Hyrenbach et al. 2000).

Several studies have quantified the overlap of satellite-tracked albatrosses with management jurisdictions, such as Exclusive Economic Zones (EEZs) and MPAs. These data provide valuable lessons about the potential use of marine zoning to protect these far-ranging species during different parts of their life cycle (see table: 'Studies of albatross use of Marine Protected Areas'). First, albatross species breeding concurrently at the same location may use marine reserves differently, if they forage in different oceanic habitats. For instance, during their incubation and brooding period two sympatrically breeding albatrosses spent approximately 31 per cent (black-browed) and 15 per cent (grey-headed) of their at-sea time within the Macquarie Island Marine Park, a 160,000-square kilometre (61,760-sq mile) protected area within the Australian EEZ (Terauds et al. 2006). Telemetry studies have also documented substantial variability in MPA use within a given species, as trip durations and foraging ranges expand and contract during the breeding season. For instance, foraging waved albatross from Española Island spend over two thirds of their time within the Galapagos Marine Reserve (GMR) during the brooding period, but the use of the reserve was reduced significantly during the incubation (15 per cent) and chick-rearing (10 per cent) periods. Finally, some species, such as the black-footed albatross, may forage within MPAs located very far from their breeding colonies. Chick-rearing birds from Tern Island, North-West Hawaiian Islands, commuted 4500 kilometres (3000 miles) to forage within National Marine Sanctuaries located in the California Current. Yet, when post-breeding birds were subsequently tagged within these sanctuaries, they spent a small amount of time within these

ABOVE Marine Protected Area (MPA) designs for core albatross habitats (A) and the 'footprint' of different potential threats (B). The conceptual match/mismatch between the protective 'buffers' and the 'footprints' allows managers to evaluate whether a specific MPA design will offer albatrosses protection from a given threat.

protected waters and ranged widely across the North Pacific. These studies underscore the need for detailed data on albatross distributions as the foundation for an integrated and comprehensive approach to the conservation of their populations and oceanic habitats (Hyrenbach et al. 2000, Gilman 2001).

Marine zoning, which seeks to manage the whole range of human activities (commercial and recreational fisheries, oil and gas exploration and drilling, maritime transportation, recreational activities, military exercises, ecotourism, aquaculture and other extractive activities such as sea mining) and the conservation of marine resources (including fisheries, protected species, and both benthic and pelagic habitats), provides a flexible framework for integrated albatross conservation.

The key to effective marine zoning lies on the ability to mitigate detrimental impacts on the natural resources, by segregating non-compatible uses in time or in space. Zoning concepts are not new, having been used in terrestrial systems for decades. The segregation of commercial and residential areas within cities, the design of highway and railway transportation corridors, and the establishment of national parks are classic examples of land-use planning. Similar large-scale zoning approaches are being advocated for the management of marine systems (Norse et al. 2005).

The marine zoning 'tool kit' includes a wide array of protective measures, with varying degrees of spatial coverage and design flexibility. Fishery monitoring (e.g. observer programmes) and bycatch mitigation measures (e.g. seabird-scaring lines or 'tori-lines') can be applied in a diffuse fashion (e.g. across entire fishing fleets), or can be focused on priority times and areas of highest potential threat to protected species. Information on albatross movements and fishing effort distributions are being used to identify 'high-risk' times and areas of high albatross overlap with fisheries (BirdLife 2004). Marine reserves and temporary fishery closures are best suited for instances when albatrosses forage within fairly restricted areas around their colonies or commute to specific foraging grounds. Yet, effective reserve designs will require an understanding of the dynamics (spatial and temporal predictability) of the oceanographic habitats exploited by albatrosses (Hyrenbach et al. 2000, 2006).

The vision and political will to develop an integrated albatross conservation plan are coming together, spurred by the development of marine zoning. A particularly novel conceptual development entails the private leasing and ownership of submerged lands, which offers many exciting possibilities for the conservation and management of marine resources, including the establishment of MPAs for the protection of marine resources and habitat restoration. Currently, local communities own and lease submerged lands for commercial fishing and pearl harvesting, national governments grant marine concessions to support growing aquaculture

Studies of albatross use of Marine Protected Areas (MPA)

Protected Area, Location	Albatross species	Period of life cycle	Time in MPA (%)	Reference
Macquarie Island Marine Park (MIMP), Australia	Black-browed (*Thalassarche melanophrys*)	Incubation/Brooding	31%	Terauds et al. 2006
	Grey-headed (*Thalassarche chrysostoma*)	Incubation/Brooding	15%	
Galapagos Marine Reserve (GMR), Ecuador	Waved (*Phoebastria irrorata*)	Incubation	10%	Anderson et al. 2003
		Brooding	68%	
		Chick rearing	15%	
US National Marine Sanctuaries, Central California (MNS), USA	Black-footed (*Phoebastria nigripes*)	Brooding	0%	Hyrenbach et al. 2006a
		Chick Rearing	11%	
		Post-breeding	19%	

and wind-power industries, and private organisations own and manage islands and reefs for conservation and ecotourism. These examples illustrate the potential application of this approach to the 'grass-roots' advancement of marine zoning.

Marine reserves designed to manage longlining and trawling on continental shelf-slope regions around albatross colonies, coupled with reductions of fishing effort through licence buy-back programmes, could protect those critical habitats where breeding albatrosses concentrate to forage. Albatross conservation during other periods of their life cycle, when they disperse widely, would require the use of the diffuse management approaches described above.

Different threats have characteristic 'footprints', which directly influence the ability of MPAs to mitigate their impacts. In fact, certain threats have such large-scale impacts that MPAs are ineffective conservation measures. For instance, whereas the main line of a pelagic longline is up to 100 kilometres (62 miles) long, oil may extend many hundreds of kilometres downstream from the site of a spill before it dissipates. Therefore, MPAs should include buffers designed to displace the 'footprint' of potential threats away from critical albatross habitats. For instance, while the classic core and buffer MPA illustrated in the diagram opposite would exclude fisheries bycatch impacts, it would not protect the core albatross habitat from an oil spill. This inability of MPAs to mitigate large-scale anthropogenic impacts with basin-wide 'footprints', such as climate change and plastic pollution, emphasises the need for a comprehensive approach to albatross conservation over entire ocean basins.

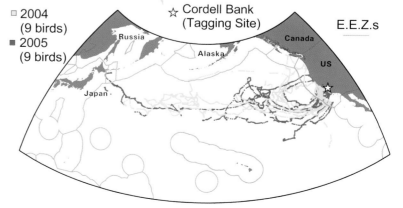

26
Conversation with an Ex-High Seas Poacher
Tui De Roy

Synopsis: During a long sea passage author Tui De Roy meets a fishing captain with a great fondness for albatrosses, who recounts his experiences of a past life aboard a pirate longliner and how he devised his own mitigation techniques to avoid killing birds.

In my time at sea I have met many old salts and spent long hours on the bridges of an eclectic variety of vessels, but the most interesting conversations have always been with fishing captains, especially those plying the high seas. These men breathe the sea and live the sea, and sense the pulse of all that lives within it — if they didn't they wouldn't be good at catching fish. When they feel like talking, the depth of their knowledge and observational skills is invariably fascinating.

One such conversation started many hundreds of miles out of South Africa. The sea was calm, with shimmering shafts of sunlight strafing the great oceanic swells. All around us petrels of many sizes and shapes flitted and banked, while several great wandering albatrosses cruised up and down the slipstream of the ship.

'Ah,' said the captain unprompted, 'I do love these birds. You know, as fishermen we are here for the money, to earn our living. But there's more to it. These great birds are really what the sea is about.' Then he frowned, shaking his head, his lilting accent adding emphasis, 'It's awful how many of them die around the fishing boats, I just can't accept that. I used to fish out there with the illegal fleets, and it's disgusting. No, I couldn't accept that, I couldn't kill those birds.'

I had always wanted to hear, in their own words, about the so-called pirate fisheries responsible for untold numbers of albatross deaths, far beyond all official control or monitoring efforts. I spent the rest of the afternoon plying him with questions.

'Please understand, I'm happy to help your project by sharing my experience, but I don't want to mention my name, because my ship was operating illegally. I poached, I was caught, I paid my penalty. I don't want to be involved any more.'

'L' was born in Portugal but now lives in South

RIGHT AND ABOVE LEFT Referring to the great Wanderers as 'the huge white ones' (such as these Antipodean albatrosses on their feeding grounds), captain 'L' remained their devout protector even while plundering the oceans, evading rules and quotas.
ABOVE RIGHT The inspired initiative of an enlightened fisher can make all the difference for white-capped albatrosses travelling from New Zealand to feed in the Indian Ocean, where they are threatened by poorly controlled fishing operations.

Africa. He has been a fisherman since he was 16 and spent three years in the late 1990s working on a Portuguese-flagged vessel pursuing Patagonian toothfish (more recently renamed Chilean seabass) by deep-sea longlining across the Indian Ocean, moving with the rest of the illicit fleet from one fishing ground to another as the catches dwindled.

Q: *'What sort of operation did you run, and how many other ships were there? What was your experience of that fishery?'*
A: 'We had no [fishing] licence at all, and we were shooting 15–20,000 hooks per day. There were dozens of ships working the area, maybe 50 or 60, maybe more, from all over, but I'm not sure what nationalities were involved because we very seldom talked. At first we sailed from Kenya, then South Africa, and when the catches got worse, from Mauritius. We fished year round, starting with the African Rise and the South African [Prince Edward] Islands, but heading east when the catches dropped, to the Kerguelen Islands, MacDonald and Crozet. Each time the fishing got worse, so we went further east. Once I was caught by a French patrol vessel named the *Albatross* in the Kerguelen Islands, and escorted all the way back to Reunion Island [3200 km (2000 miles)]. We were held for six months, along with five or six other ships from various countries — Portugal, Spain, Panama and others I can't remember. When they gave us back our passports I came home to Cape Town.'

Q: *'What is it like being a poacher, what was your role? Tell me about the albatrosses you encountered.'*
A: 'When you poach you don't care, you just up and shoot, up and shoot. But for me, myself, it was different, because I like to protect the birds, especially the albatross. I was both the Captain and the Fishing Master, so I was the one catching the fish. There were lots of albatrosses around us all the time, mainly the small dark ones, the big white ones [the Wanderers] always stayed further away.'

Q: *'How did you avoid catching albatrosses, and why?'*
A: 'We used two lines with 25 to 30 flags attached, about 35 metres [115 ft] long, which I made on board myself to keep the albatrosses from getting at the sinking baits. When we'd shoot [the longlines], we put these lines out astern, one on each side of the ship, trailing far behind. I used the flags every time, because as a fisherman I don't like to catch the birds, because life at sea is about our life and theirs.'

Q: *'Did the other vessels use the same technique? Do you think they know the effect they're having, and do they care?'*
A: 'I never saw another boat using flags. I really don't know if the other captains didn't care, or they just don't understand; they have no experience [avoiding birds]. But I read it in the book — or is it a magazine? — *Fishing World*, from England. In my time we caught maybe one or two albatrosses per week — only the small ones, never the big ones — but that is only because I used the flags, and because I shot the lines at night. Yes, the *Fishing World* is how I saw that something could be done, and so I did it, and it works — yes, it does work.'

For a long time afterwards we were silent, just watching the birds going about their business along the wavetops. Here was a man who, without ado, had taken the matter of keeping albatrosses from falling victim to high seas fisheries into his own hands, even while he dodged the laws of the countries whose waters he worked. His story confirmed that the work of BirdLife International, Southern Seabirds Solutions, and the many others who strive to inform and inspire fishing captains around the world is the way forward to save the albatross.

ABOVE On a totally windless day, a black-browed albatross skims the unusually waveless Drake Passage between Antarctica and South America, a region feared for its vicious storms. According to the fishing captain, these birds were the most persistent ship followers, which he took pains to protect.

Part Three
Species Profiles
Julian Fitter

Introduction to Albatrosses, Mollymawks and Gooneys

The albatross family (Diomedeidae) is divided into four genetic groups:
- The great albatrosses of the genus *Diomedea*: 6 species.
- The mollymawks of the genus *Thalassarche*: 10 species.
- The sooty albatrosses of the genus *Phoebetria*: 2 species.
- The northern albatrosses, or gooneys, of the genus *Phoebastria*: 4 species.

Etymology

The word albatross is derived from the Greek *kados* or 'bucket', via the Arabic *al cadous* or *quadras* and then through the Portuguese *alcatruz* or Spanish *alcatraz* meaning 'pelican' due to its capacious bucket-like bill pouch. Due no doubt to the size, the name was transferred and then evolved to albatross, but does not refer to the whiteness of the plumage. The name mollymawk is an old sailor's term for albatross, particularly the smaller black-winged ones. The origin of this word is thought to be a corruption of the Dutch *mallemugge* or *mallemuck*, or alternatively the French *molle-muc*, all of which translate as 'foolish gnat', a name first given to the northern fulmar, *Fulmarus galcialis*, from its habit of swarming behind fishing boats. We have used the term in this book to describe the genus *Thalassarche*.

Gony, goney, gonia, goony or gooney is the term used in some publications and in colloquial speech to describe the three North Pacific albatrosses of the genus *Phoebastria*, the short-tailed, Laysan and the black-footed. Originally the word referred to any albatross and is of similar origin to mollymawk. Gony is an old English dialect word for simpleton or booby and was used by whalers who expanded its usage. The term 'goney' is still used on Tristan da Cunha in reference to the Tristan albatross. The local Japanese name for the short-tailed albatross is *ahodori* or 'fool bird'.

Both the names mollymawk and gooney apparently refer to their perceived foolishness in not fleeing from humans when attacked or having their egg stolen, a behaviour stemming from the fact they had no natural land predators before the arrival of humans. They are also, in contrast to their effortless elegance in the air, rather ungainly on land, and their rolling head-down gait can appear clown-like.

Nomenclature and taxonomy

Until quite recently it was accepted that there were 13 species of albatross. With advances in the science of genetics, a new taxonomy has developed, based on molecular (genetic) characteristics, which has identified 24 species. Not all experts and organisations involved accept all of these species so for the purposes of this book we have adopted the nomenclature used by BirdLife International, which recognises 22 species only.

The great albatrosses

The genus *Diomedea* is named after Diomedes, King of Argos who fought with the Greeks in the Trojan War. Diomedes offended the goddess Athena and was punished by having some of his followers turned into large white birds that were both gentle and virtuous. An alternative version has them metamorphosing into birds on Diomedes' death. The term 'birds of Diomedes' was originally applied to the Yelkouan or Mediterranean shearwater, *Puffinus yelkouan*.

The great albatrosses are the largest of all seabirds and have the longest wingspan of any living bird, sometimes exceeding 3.5 metres (11.5 ft). They are found exclusively in the southern oceans and have impressed and inspired poets, artists, sailors and other travellers for generations. A juvenile wandering albatross is very likely the inspiration behind Coleridge's *Rime of the Ancient Mariner*, which turned the word 'albatross' into a figure of speech.

Until recently these very large albatrosses were separated into just two species, the wandering and the royal (Nelson 1980). However, with the aid of DNA testing Robertson and Nunn (1998) divided them into seven individual species: five Wanderers (Antipodean, Gibson's, Tristan, Amsterdam and wandering, the last sometimes also called the snowy albatross) and two royals (northern and southern). These changes are widely accepted, apart from Gibson's, *Diomedea gibsoni*, which is now regarded as only a subspecies of the Antipodean, *D. antipodensis*.

As a result the name 'wandering albatross' can refer both to the individual species *Diomedea exulans* and the group of four similar species, wandering, Amsterdam, Tristan and Antipodean, that were originally considered as one. To avoid confusion we have distinguished between the two by using a capital 'W' when referring to the group, as in the 'Wanderers' and a lower case 'w' when referring to the now-recognised individual species, wandering albatross, *D. exulans*.

Their plumage varies from almost all white to almost all brown, with every possible variation in between. Because of this, identifying the individual species at sea can be difficult unless you are in the vicinity of their breeding grounds. The ratio of light and dark colouring changes with age, becoming gradually whiter, which makes it very difficult to distinguish males from

ANTIPODEAN ALBATROSS

PREVIOUS PAGE New arrivals swell the ranks of gamming southern royal albatrosses near Mount Lyall on Campbell Island.

186 • Species Profiles

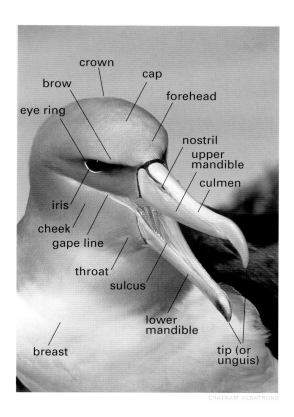
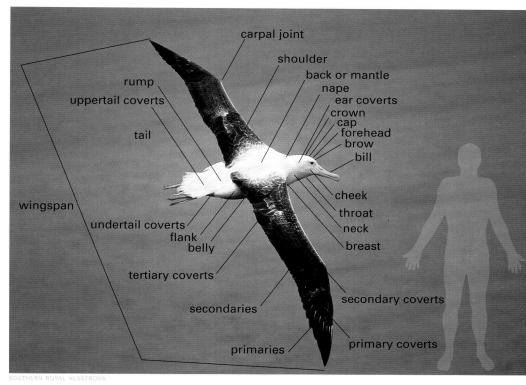

ABOVE The main parts of an albatross as used in the species descriptions on the following pages. The human figure gives an indication of scale.

females and adults from juveniles. This helps to explain why the current six or seven species were previously regarded as just two. To aid offshore identification, J.D. Gibson devised the Gibson Plumage Index. This was refined by Jouventin and others in 1989, and gives a very good indication of the variations involved and the complexities of identification at sea.

The mollymawks

Before the advent of DNA analysis, there were thought to be just five species of mollymawks, but Robertson and Nunn (1998) split them into 11 species, of which 10 have been generally accepted. Thus the shy was divided into four — Salvin's (*Thalassarche salvini*), white-capped (*T. steadi*), shy (*T. cauta*) and Chatham (*T. eremita*); the black-browed into Campbell (*T. impavida*) and black-browed (*T. melanophrys*); and the yellow-nosed into Atlantic (*T. chlororhynchos*) and Indian (*T. carteri*). The only one that remained unchanged was the grey-headed albatross, *T. chrysostoma*, whereas the splitting of Buller's into a southern and northern species (the latter sometimes referred to as the Pacific albatross, *T. platei*) was rejected.

As a genus the mollymawks are relatively easy to identify at sea. They are very large, straight-winged birds that glide effortlessly close to the surface. They are larger than any other seabird apart from the great albatrosses, which are significantly larger still. They are all very similar in size and shape, and their plumage, while not uniform, conforms to a basic template: mostly white body and underwings, pale head, dark upper wings, mantle and tail, with white rump. In general terms they are divided into two size groups, with the shy tribe being slightly larger that the others, though in reality this distinction is not readily noticeable at sea unless they are flying together. The only other species of a similar size, but not colour scheme, within their range are the two giant petrels.

Population estimates

In compiling this part of the book it has become obvious that obtaining clear and accurate data on actual population numbers is very difficult. The only effective means for calculating an overall population is to use the number of breeding pairs as a basis, yet even this is fraught with problems.

Firstly, counting nests is not easy as many species breed in very isolated areas and, while some colonies are closely packed, others are very spread out and often among obscuring vegetation.

Secondly, while we may know that some species breed yearly whereas others do so every other year, this is never 100 per cent true and a number of pairs (in some cases up to 35 per cent) will skip a season at random. As a result we will rarely have a fully accurate figure for the total breeding population.

Thirdly, all albatrosses spend several years at sea before breeding, a period which may vary from as short as six or seven years to as long as 22. This means a very large population of immatures is likely to

Species Profiles • 187

CAMPBELL ALBATROSS

WANDERING ALBATROSS

Range

The great albatrosses are regarded as the ultimate travellers, covering enormous distances at sea. In 1992 a female wandering albatross, *Diomedea exulans*, tracked from South Georgia, was found to have covered 25,075 kilometres (15,571 miles) in 36.3 days, an average of 690 kilometres (429 miles) per day. In one 3.5-day period she travelled 4447 kilometres (2761 miles) at an average speed of 53.6 kilometres per hour (33 mph). She was still travelling towards the traditional wintering grounds off the west coast of Australia when her transmitter failed. When not breeding, northern royal albatrosses regularly cruise right around the world, as do the much smaller grey-headed albatross, one of which even flew twice around in its year off. However, we still have a fairly incomplete picture; indications of range with each species provide only general guidance. The range of adults during the breeding season is normally appreciably reduced, particularly in the early period after hatching, when chick feeding is quite frequent.

Longevity

While accurate data for most species is not available, all albatrosses are very long-lived birds indeed. The oldest on record was a female northern royal albatross named 'Grandma' at Taiaroa Head on the South Island of New Zealand, estimated to be at least 61 years old when last observed in June 1989. For the smaller albatrosses, or mollymawks, the oldest known individual was a female Buller's, *Thalassarche bulleri*, last recorded in 1994 when she was at least 52 years old.

Breeding

All albatrosses lay only one egg per breeding season, which they will not replace that year if it is lost. Both parents share incubation, in shifts lasting up to two weeks or more, and are equally responsible for chick brooding and provisioning until it becomes independent, with the male generally doing slightly more than the female in the ratio of 55 to 45. Chicks generally weigh significantly more than adults prior to leaving the colony.

A universal requirement for the family are oceanic islands, devoid of land predators, on which to breed, these often being quite remote and often inaccessible, and in some cases quite distant from their primary feeding grounds. The smaller albatrosses of the mollymawk group (*Thalassarche*), and to a lesser extent the two sooties (*Phoebetria*), build sturdy pedestal nests from a mortar of mixed mud, vegetation, feathers and anything else available, painstakingly patted together into a hard, cone-shaped structure with a deep bowl at the top, sometimes acquiring the look and texture of a terracotta vase. These nests are often reused annually. The larger species of the genus *Diomedea* build a similar structure, but broader and looser, and made of more cushiony materials, such as grass

exist in addition to the adult breeding population, consisting of both actual breeders and those taking a holiday in any given year.

As an example, take a breeding population of 1000 pairs, then assume that they have a 50 per cent success rate, thus producing 500 chicks that fledge each year. Further assume that these chicks start to breed when 10 years old, but that only 50 per cent of them survive to this age, and that the death rate is evenly spread over that 10-year period. The result of this exercise shows that for every 1000 pairs, there are 2500 non-breeders. While these assumptions are not unreasonable, the factors will be different for each year and for each breeding population and each species, turning any estimate more into a 'guesstimate', and one that may well be so inaccurate as to cause problems in advancing the conservation of the species.

We have consequently not ventured to give total population figures at all, refraining from the use of extrapolated totals in favour of known or estimated annual breeding populations only.

The most important considerations are population trends and relative numbers of breeding pairs over time, as indicators of whether a species is increasing, decreasing or stable. Conservation efforts can then be concentrated where most needed.

and moss. The northern albatrosses (*Phoebastria*) use mainly hollows scraped out in the sand and surrounded with vegetation debris, with the notable exception of the waved albatross, which uses no nest at all. Breeding locations for the southern species vary latitudinally from Tristan da Cunha at only 37 degrees south to Diego Ramirez (Chile) at 56 degrees south. The northern species are found only in the Pacific Ocean.

The great albatrosses (royals and Wanderers) all breed biennially, eggs being laid from mid-November through mid-March depending upon the species and breeding location, with chicks fledging 11–12 months later. This means that for successful breeders the end of one season very nearly overlaps with the beginning of the next so they must take a year off in between in order to regain condition.

Of the smaller species, only the grey-headed and the two sooties also breed biennially, with all others nesting on an annual basis. Pair-bonds are generally permanent, but if a mate dies, the survivor is likely to find another mate. Immatures often form temporary pairings prior to actually breeding.

Courtship

Courtship is an elaborate and poorly understood ritual, which has many elements and variations. Wareham (1996) lists a dozen distinct elements of the nuptial display of the great albatrosses (*Diomedea*). These include: billing, bowing, yapping, bill snaps, gawky look, head shake, sky call, wing stretch and sway walk. Accompanying these visual displays are a remarkable assortment of vocalisations, moos, whistles, grunts, groans, wails and cackles. These very visual and vocal sessions are often interspersed with bouts of mutual preening. Aerial display is also used, involving the 'sky call' in flight, generally while flying low over the colony, and in conjunction with ground display.

Mollymawk (*Thalassarche*) courtship is similar in some respects to the courtship of the great albatrosses, but slightly more restrained. With some species, such as the black-browed albatross, this is likely due to their nesting very close together, which precludes elaborate dances and displays. In most species courtship involves bowing, often including fanning out the tail, mutual preening, bill fencing, gaping and vocalising, including loud monosyllabic wails, croaks, rattles and a rather harsh descending laughing call. This may be accompanied by bill clapping and exposing the vividly coloured gape line, a narrow rumple of skin that runs from the base of the bill across the cheek, hidden beneath the feathers at other times. This last behaviour can happen at any time of the year and is not limited to the period immediately prior to mating. After copulation the female leaves the breeding area for a period which can be up to 20 days, returning to the site to lay the single egg.

Perhaps the most complex courtship rituals are to be found among the northern albatrosses (*Phoebastria*) who follow intricately choreographed dance routines involving precise sequences of gestures such as sky-pointing, fake-preening, shoulder-hiking, tiptoeing and pirouetting around one another, all performed at a rather frantic pace.

In the sooties (*Phoebetria*) courtship is primarily performed on the wing, with pairs flying together in locked formation mirroring each other's moves. This is interspersed with more subtle yet prolonged ground sessions at the chosen nest site, involving nodding, high-stepping, tail fanning and twisting, and bill snapping. Additionally, the haunting 'pee-oow' advertising call, performed by one bird sitting on a ledge in response to another passing overhead, is legendary.

Much albatross courtship display behaviour is performed by non-breeders and juveniles who appear to be practising for the future. Depending on the species, these may involve up to a dozen individuals, and is referred to as 'gamming', a term used by sailors meaning a social visit or friendly interchange.

In addition to nuptial displays, great albatrosses have a number of aggression and defence displays, including defence billing, bill clop, bill clapper and charge. Actual fighting is very rare except at sea when scavenging. Smaller species, on the other hand, may engage in vicious fights, whether over mates, stolen nesting mud or just over space in tightly packed colonies, such as in the colonies of the Salvin's albatross. At the other extreme the sooties appear to be totally non-aggressive.

Albatrosses: Status, distinctive or diagnostic features, breeding and range

This quick-reference table is based on normal, mature plumage. The bold place names in the 'Breeding' column indicate the most important breeding sites for that species, along with abbreviations for country: South Africa (SA), France (Fr), United Kingdom (UK), New Zealand (NZ), Australia (Au).

Species	Appearance	Other features	Breeding	Range
WANDERERS (*Diomedea*)				
Wandering *D. exulans* Vulnerable	All white apart from wingtips and trailing edge of wing. Frequent pink stain on ear coverts. Whitest of great albatrosses. Closest to southern royal.	Largest of Wanderers. Bill Pink with dull yellow tip.	**Prince Edward** (SA), **Crozet, Kerguelen** (Fr), **Bird** (South Georgia, UK), Macquarie (Au)	Wide-ranging, all southern oceans.
Amsterdam *D. amsterdamensis* Critically endangered	Darkest of great albatrosses. White belly and face. Closest to Antipodean, but darker, and distinct ranges.	Black line on cutting edge of upper mandible (as with royals). Size similar to Tristan.	**Amsterdam Island** (Fr)	Mainly Indian Ocean.
Tristan *D. dabbenena* Critically Endangered	Largely white body, mantle and tail. Dark upper wing. Striking vermiculate plumage especially on breast and neck.	Size similar to Amsterdam, but lighter plumage. Bill pink with dull yellow tip, no dark cutting edge.	**Gough, Inaccessible** (Tristan group, UK)	Mainly South Atlantic but occasionally east to Australia.
Antipodean *D. antipodensis* Vulnerable	White body with brown vermiculations, white belly and face. Adult females remain noticeably brown with white face.	Second largest of Wanderers. Pink bill with dull yellow tip.	**Antipodes; Adams,** Campbell, Disappointment and Auckland in Auckland group (NZ)	Largely South Pacific, Australia to South America.
ROYALS				
Northern royal *D. sanfordi* Endangered	All white body mantle and tail. Upper wing all dark in adults. Dark on underwing extends to carpal joint. Lacks pink ear coverts of most Wanderers.	Black line on cutting edges. Bill pink. Smaller than southern royal, similar to wandering.	**The Sisters, The Forty-Fours** (Chatham Islands), Taiaroa Head (NZ)	Circumpolar, esp. Patagonian Shelf; infrequent in Antarctic waters.
Southern royal *D. epomophora* Vulnerable	All white body, mantle and tail. Upper wing whiter that northern royal. White leading edge to inner wing. Lacks pink ear coverts of most Wanderers.	Largest albatross, longest wingspan. Pink bill with black line on cutting edges.	**Campbell**; Adams, Enderby and Auckland in Auckland group (NZ)	Circumpolar but not into Antarctic waters. East coast of South America.
MOLLYMAWKS (*Thalassarche*)				
Black-browed *T. melanophrys* Endangered	Head white. Distinctive dark eye patch. Underside of wing white with broad dark leading edge and narrower trailing edge.	Bright orange-yellow bill. Dark eye.	**Falklands, South Georgia** (UK); **Diego Ramirez, Ildefonso, Diego de Almagro** (Chile); Heard, MacDonald, Macquarie (Au); Campbell, Antipodes, Crozet, **Kerguelen** (Fr)	Circumpolar, but mainly Atlantic and Indian Oceans.
Campbell *T. impavida* Vulnerable	Dark eye-patch, underside of wing has less white than black-brow.	Bright orange-yellow bill. Pale yellow eye (iris).	**Jeanette Marie and Campbell** in Campbell group (NZ)	Southern Ocean south of NZ and Australia.
Buller's *T. bulleri* Near Threatened	Grey head with paler forehead and crown. Broad dark leading edge to underwing and narrower trailing edge.	Bill black with bright orange-yellow upper and lower ridges.	**The Forty-Fours, The Sisters** (Chatham group); **Snares, Solander** & Three Kings (NZ)	South Pacific, Australia to South America, up to Peru.
Shy *T. cauta* Near Threatened	Very pale grey head with white crown. Dark triangular eye patch.	Bill dull yellow with dark yellow tip.	Pedra Branca, **Mewstone, Albatross** (Tasmania, Au).	South Indian Ocean and Tasman Sea to South Africa.

Species	Appearance	Other features	Breeding	Range
White-capped *T. steadi* Near Threatened	Distinct white crown. Slight greying on cheeks. Dark triangular eye patch.	Bill dull yellow with dark yellow tip.	Bollons in Antipodes group; **Disappointment**, Adams and Auckland in Auckland group (NZ)	Circumpolar, east coast of South America; South Africa.
Salvin's *T. salvini* Vulnerable	Mid-grey head. Black thumb mark at base of leading edge of underwing.	Bill dull horn with cream upper ridge, yellower at base, black smudge on either side of lower mandible tip.	**Bounty**, Snares (NZ)	Indian and Pacific Oceans, esp. South Africa.
Chatham *T. eremita* Critically Endangered	Dark grey head with dark eye patch. Black thumb mark at base of leading edge of underwing.	Bill chrome-yellow with black tip to lower mandible.	**The Pyramid** (Chatham Islands) NZ.	South Pacific, Tasmania to Peru.
Grey-headed *T. chrysostoma* Vulnerable	Dark grey head with pale forehead. Broad dark leading and trailing edges to underwing.	Bill black with orange-yellow upper and lower ridges.	**Bird & Willis** (South Georgia); **Diego Ramirez** (Chile), **Marion**, Prince Edward (SA), **Crozet**, **Kerguelen** (Fr), **Campbell** (NZ), Macquarie (Au).	Circumpolar between 39° and 64°S.
Atlantic yellow-nosed *T. chlororhynchos* Endangered	Pale grey head, broad dark leading edge on underside of wing.	Bill black with bright orange-yellow upper ridge ending in a 'U', orange tip.	**Tristan da Cunha, Nightingale**, Inaccessible, **Gough**. (Tristan group, UK)	South Atlantic between 15° and 45°S.
Indian yellow-nosed *T. carteri* Endangered	White head, broad dark leading edge to underside of wing.	Bill black with bright orange-yellow upper ridge ending in a 'V', orange tip.	**Prince Edward** (SA), **Pingouins**, **Apotres** (Crozet), Kerguelen, **St Paul's**, **Amsterdam** (Fr)	South Indian Ocean, South Africa and east to New Zealand.
SOOTIES (*Phoebetria*)				
Sooty *P. fusca* Endangered	Entirely dark plumage. Narrow diamond-shaped tail.	Slim and streamlined. Sulcus stripe on bill yellow. Aerial courtship.	**Gough**, Inaccessible, Nightingale and **Tristan** (Tristan group, UK); Prince Edward; **Marion** (SA); Crozet, Amsterdam (Fr)	South Atlantic and Indian Oceans north to 20°S.
Light-mantled *P. palpebrata* Near Threatened	Dark plumage apart from pale grey mantle and flanks. Narrow diamond-shaped tail.	Sulcus stripe on bill pale blue. Slim and streamlined. Aerial courtship.	Pr. Edward, Marion (SA); **S. Georgia** (UK), **Crozet**, **Kerguelen** (Fr); Heard, Macquarie (Au); **Auckland**, **Campbell**, **Antipodes** (NZ)	Circumpolar, south to pack ice edge.
NORTHERN (*Phoebastria*)				
Waved *P. irrorata* Critically Endangered	Brown upper wing, paler underwing. Fine barring on breast and abdomen. Yellow wash on head.	Very large yellow bill. Distinctive eyebrows.	**Española** (Galapagos Islands); Isla La Plata (Ecuador)	Eastern tropical Pacific, Galapagos to Peru.
Short-tailed *P. albatrus* Vulnerable	Black and white upper wing, white body, yellow and rust on head and neck, underwing white, tail black.	Bill pink with bluish tip.	**Torishima**, Senkaku-retto (Japan)	North Pacific, tropics to Aleutians, mainly continental shelf.
Laysan *P. immutabilis* Vulnerable	White body with dark upper wing and mantle, tail black.	Pinkish yellow bill with dark grey tip.	**Laysan, Midway Atoll** and several other Hawaiian Islands (US); Bonin (Japan). Guadalupe; Clarion & San Benedicto (Mexico)	North Pacific north of 20°N, especially north of Hawaiian islands.
Black-footed *P. nigripes* Endangered	All dark plumage with white area around bill.	Black bill and feet.	**Midway Atoll, Laysan** & other Hawaiian Islands (US); Torishima, Senkaku-retto and Bonin (Japan)	North Pacific north of 20°N, Japan to California.

Wandering Albatross *Diomedea exulans*

Alternative or previous names: Snowy albatross, Cape sheep.
First described: Linnaeus, 1758. *Systema Naturae*, ed.10, 1:132.
Taxonomic source: Brooke (2004), Robertson and Nunn (1998).
Taxonomic note: *Diomedea exulans* (Sibley and Monroe 1990, 1993) has been split into *D. exulans*, *D. dabbenena* and *D. antipodensis* following Brooke (2004) and contra Robertson and Nunn (1998) who also split *D. antipodensis* into *D. antipodensis* and *D. gibsoni*.
Origin of name: From Latin *exsulans*, 'living as an exile'. This and the common name both refer to its high-seas roaming.
Conservation status: Vulnerable.
Justification: Restricted breeding range. Populations at two colonies have declined by some 50 per cent over 70 years, while overall the species appears to have declined by some 30 per cent over the same period. The population may now have stabilised, at least at some colonies.

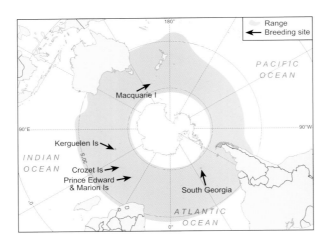

Description
The largest of the four species of Wanderers; older adult males develop an almost all white plumage, hence its alternative name of 'snowy' albatross.
Identification: Very large, very white seabird, similar only to royal albatrosses (which always retain more black on the upper wing than mature adult male wandering albatrosses). At close range, pink ear coverts and the lack of a black edge to upper mandible are strong distinguishing features. Juveniles hard to tell from other Wanderers, except possibly by size.

Adult male
Body and rump: White, mantle also white, but often with fine grey vermiculation on neck and shoulders, depending on age.
Head and neck: White, often with salmon pink 'stain' on ear coverts, mainly in breeding birds. The nature of this colouration is unknown.
Bill: Massive, flesh coloured with pronounced pale horn-coloured hooked tip.
Wings and tail: Wing largely white (both upper and under), but with black tip and black trailing edge, plus dark primary coverts on upper wing. Tail white with dark tip, which gradually whitens from centre outwards, but always retains black tips to outer feathers.

Adult female
Generally less white than male; mantle and wings, especially secondaries and coverts, always retain some dark mottling.

Juvenile
Dark brown head and body, with white side of head and throat giving a hooded appearance. Wings dark greyish brown, underwing white with dark tips and trailing edge. Bill similar to adult. Paler plumage develops gradually over many years, no details are available on minimum age for whitest plumage.

Size
The largest of the Wanderers; males generally larger than females.
Body length: 110–135 cm.
Wingspan: 2.5–3.50 m.
Weight: Male 8.2–11.9 kg. Female 6.7–8.7 kg.
Bill: Male 163–180 mm. Female 155–171 mm.
Egg: Length 114–142 mm. Diameter 78–86 mm. Weight 450–530 g.

Population and distribution
Estimated to be some 8500 annual breeding pairs at the four main breeding sites: South Georgia (mainly Bird Island), Prince Edward and Marion, Crozet (mainly Ile aux Cochons) and Kerguelen, plus a tiny cluster of less than a dozen pairs on Macquarie.

Oceanic range
Circumpolar; travels throughout the southern oceans, especially immatures and non-breeders. While breeding, males tend to forage further south than females. Adult males are 20 per cent heavier than adult females, but their wing area is only 7 per cent greater, giving them a 12 per cent greater loading. The implication is that males are better adapted to the stronger winds of the higher latitudes.

Breeding
Colonies spread out over a considerable area in what is best described as a loose association on flat or gently sloping land, or on ridges. Windward coasts are generally most favoured. Individuals

may breed as early as year seven, but 10 years is more normal. Pairs breed biennially, though 30 per cent of successful breeders and 35 per cent of failed breeders defer breeding by a year. They arrive at the breeding grounds from mid-December through January, the timing varying between island groups.
Courtship: An elaborate ritual similar to all the great albatrosses, involving outstretched wings, head waving and lifting, bill clapping, tail cocking; all while emitting a remarkable variety of sounds including neighs, groans, wails, croaks and trills.
Laying: A single large white egg, with occasional reddish blotches at larger end, laid from mid-December through mid-January.
Incubation: 75–83 days, hatching in March and early April.
Fledging: Chick is downy white at hatching, which may take three to four days. Brooded by both adults for around 30 days, with feeding taking place every two to four days, becoming gradually less frequent as the chick grows. Fledging takes a total of 263–303 days on South Georgia, six to seven days less on Crozet, departing in late November or December.
Breeding success: Around 90 per cent but this varies with location and from year to year dependent on food supply and weather. Survival of fledged young to breeding age is around 30–35 per cent.

Food

Fish, as well as squid and other cephalopods, but rarely crustaceans, taken from the surface or occasionally by shallow lunge diving to 2 metres (6.5 ft) depth. Scavenges around fishing boats or dead marine life for carrion.

Threats

Fishing: Well known to scavenge around fishing boats and to be at risk when no mitigation measures are used. Changes in fishing practice around South Georgia, including the night setting of longlines, appear to have reduced the fisheries bycatch problem for that population. Commercial fishing is, however, still a significant threat to the species.
Predation: No known predators, though skuas may take an unattended egg or young chick.
Habitat loss: In some colonies, namely at South Georgia, the recolonisation by fur seals, due to population recovery after the end of sealing, has encroached heavily into nesting areas, causing birds to relocate to higher ground.

WANDERING ALBATROSS, SOUTH GEORGIA

Antipodean Albatross *Diomedea antipodensis*

Alternative or previous names: Wandering albatross, New Zealand albatross, *Diomedea exulans*, Gibson's albatross, *D. gibsoni*.
First described: Robertson and Warham, 1992.
Taxonomic source: Brooke (2004).
Taxonomic note: *Diomedea exulans* (Sibley and Monroe 1990, 1993) has been split into *D. exulans*, *D. dabbenena*, and *D. antipodensis* following Brooke (2004), but contra Robertson and Nunn (1998) who also split
D. antipodensis into *D. antipodensis* and *D. gibsoni*.
Origin of name: Antipodes Island is the breeding ground of one population. Gibson's refers to J.D. 'Doug' Gibson who developed the Gibson Plumage Index of colour variation for the wandering albatross complex.
Conservation status: Vulnerable.
Justification: The species breeds only on Antipodes Island and the Auckland Islands with a few pairs on Campbell Island, all within the New Zealand Subantarctic. There is evidence to suggest a major decline of the Auckland Island population since the mid-1980s, and the species is known to have suffered significant mortality from commercial fisheries.

ANTIPODEAN ALBATROSS, ANTIPODES ISLANDS

ANTIPODEAN ALBATROSS, ANTIPODES ISLAND

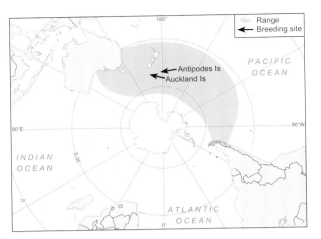

Description
Endemic to New Zealand, the Antipodean albatross is one of the four species of Wanderers, and is smaller than *Diomedea exulans*. The plumage is generally darker and browner than all but the Amsterdam albatross. Sexual dimorphism in plumage is more noticeable than with any of the other great albatrosses except the Tristan albatross. The subspecies, Gibson's, *D. a. gibsoni*, which breeds on Auckland Island, is hard to distinguish, though older Gibson's tend to be rather whiter than those from Antipodes, with breeding females often lacking the brown breast band of the latter.
Identification: Smaller and darker than the wandering albatross. Whiter body and mantle of male distinguish it from Amsterdam albatross. Female very similar to Amsterdam, also to Tristan, though generally darker. Juveniles similar to all other juvenile Wanderers.

Adult male
Body and rump: Largely white but with mid-brown vermiculations, especially on breast and mantle.
Head and neck: Face white, with brown vermiculations on neck and collar, cap brown; pink ear coverts especially on older whiter individuals.
Bill: Large, pink, strongly hooked, with dull yellow tip.
Wings and tail: Upper wing dark black-brown with some white flecking, especially near mantle. Underwing white with dark tip and trailing edge. Tail dark black-brown.

Adult female
Much browner than male, retaining hooded appearance even in mature individuals. Body, especially breast and belly, lighter than immature with white showing through the brown.

Juvenile
Dark brown, apart from white cheeks and under-wing. Underwing has dark trailing edge and tips. Bill dull grey.

Size
Males generally larger than females. Adult of subspecies *D. a. gibsoni* on average very slightly larger than *D. antipodensis*.
Body length: 110–115 cm.
Wingspan: 3.0–3.25 m.
Weight: Male 6.6–8.3 kg. Female 5.1–6.5 kg.
Bill: Male 147–155 mm. Female 139–147 mm.
Egg: Length 125 mm. Diameter 78 mm. Weight c. 450 g.

Population and distribution
Two distinct populations, with some 5000 annual breeding pairs on Antipodes and a somewhat larger number in the Auckland Islands group (mainly on Adams, but also Disappointment and a few dozen on Auckland Island itself), bringing the overall total to 11,000 annual breeding pairs. There is a population of less than 10 pairs on Campbell Island corresponding to the Antipodes Island stock.

ANTIPODEAN ALBATROSS, ANTIPODES ISLAND

ANTIPODEAN ALBATROSS, ADAMS ISLAND

Current breeding population levels appear to be relatively stable.

Oceanic range
The foraging ranges of the two island populations are significantly different. From the Auckland Islands, breeding birds forage mainly in the Tasman Sea, although some feed up the east coast of the South Island, New Zealand; non-breeders and immatures also forage in the Tasman Sea with some occurring in waters south of Australia to the south-west Indian Ocean. Breeding birds from the Antipodes occur east of New Zealand with some making extended flights to Chilean and Antarctic waters when feeding their chick; non-breeders and immatures occur east of New Zealand and across the South Pacific to South America.

Breeding
Nests in open grass and tussock land, building a large, low nest mound out of available vegetation with some soil or mud used to hold it together. There is some indication that failed breeders returning the following year are more likely to use the same nest than successful breeders who have a year off. First breeding probably 10–12 years, in line with other Wanderers.
Courtship: Similar to that of the other Wanderers.
Laying: A single large white egg laid between late December and early February on Auckland Islands, slightly later on Antipodes Island.
Incubation: 80 days, hatching in mid-March to late April.
Fledging: Fledging after 240 days, in November and December.

Breeding success: Some 65–70 per cent of eggs laid fledge. Data lacking.

Food
Largely squid, particularly giant cuttlefish, *Sepia apama*, during winter, also other cephalopods and some fish and carrion; no record of crustaceans. Food taken from the surface or within a metre or so.

Threats
Fishing: One of the species known to be at risk by longline fishing activities. With mitigation methods more frequently used in New Zealand waters, the main threats remain in international waters by ships that do not apply such measures.
Predation: Potentially pigs could affect the few dozen pairs nesting on Auckland Island, while subantarctic skuas may prey on eggs and small chicks.

Tristan Albatross *Diomedea dabbenena*

Alternative or previous names: Wandering albatross, Gough albatross, goney, gonia or gooney bird (local). *Diomedea exulans*.
First described: Matthews, 1929.
Taxonomic source: Brooke (2004); Robertson and Nunn (1998).
Taxonomic note: *Diomedea exulans* (Sibley and Munroe 1990, 1993) has been split into *D. exulans*, *D. dabbenena* and *D. antipodensis* following Brooke (2004).
Origin of name: Common name after the Tristan da Cunha group where it breeds. Latin after Roberto Dabbene, an Italian-Argentine ornithologist who first proposed the Tristan albatross as a distinct subspecies in 1926.
Conservation status: Critically Endangered.
Justification: A small breeding population almost totally confined to a single island, Gough Island, some 200 kilometres (125 miles) south-east of Tristan da Cunha in the South Atlantic. There are indications of a decline of some 79 per cent over 70 years. Breeding success is affected by the introduced house mouse, *Mus musculus*, which attacks large chicks. Endemic to the Tristan group.

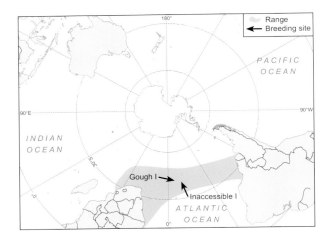

Description
The most genetically distinct of the four Wanderers, with the adult males always retaining an element of brown vermiculations on the mantle. Greatest sexual dimorphism of any albatross, with females overall much darker than males.

Identification: The smallest of all Wanderers, this species is overall darker than the wandering albatross but whiter than either Antipodean or Amsterdam, especially in males. Most easily confused with the Auckland Island subspecies of the Antipodean (Gibson's), although probably little or no overlap in range exists. Identity can only be established with certainty at, or close to, its breeding grounds.

Adult male
Body and rump: White, with rump, mantle and breast always retaining some brown vermiculations.

Head and neck: Generally white but always retaining some brown vermiculations on neck and cap.
Bill: Large, flesh-coloured with pronounced hooked dull yellow tip.
Wings and tail: Upper wing largely black though white on secondary and tertiary coverts gives mottled effect on inner wing. Underwing white with dark tip and trailing edge. Tail black.

Adult female
Very similar but wings lack white mottling, while vermiculations on body and cap overall much more pronounced and blotchy, retaining dark tendencies of juveniles throughout life.

Juvenile
Brown body with white face and throat. Upper wing brown with slight mottling; underwing similar to adult. Intermediate phases can give an overall brown mottled appearance.

Size
Males somewhat larger than females. Limited data available.
Body length: 110 cm.
Wingspan: 3.00 m.
Weight: 6.5–7.5 kg.
Bill: Male 150 mm. Female 144 mm.
Egg: Length 115–139 mm. Diameter 71–82 mm.

Population and distribution
Breeding only in the Tristan da Cunha group, the population on Gough is thought to number some 1500–2400 annual breeding pairs. Formerly present on Tristan itself, it no longer breeds due to a history of human predation, with the last recorded breeding in 1907. Recently, a few adults seen soaring over the high slopes of the volcano have raised hopes of a potential return. A remnant population on Inaccessible Island has remained stable at two or three pairs for some 50 years.

Oceanic range
Mainly South Atlantic between 23 and 45 degrees south, commonly feeding off South Africa and Brazil, with possible dispersal across the Indian Ocean as far as Australia.

Breeding
On Gough nesting is between 300 and 500 metres (985–1640 ft) elevation, mostly on the western side of the island which has more open and marshy topography. Adults return to breeding grounds in December. Nest is a low mound of grass and other vegetation, rarely more than 15 cm in height, located in open areas of rough grass and tussock. Individuals start breeding on average after 9–10 years.
Courtship: Similar to the other great albatrosses.
Laying: A single white egg laid mid to late January.
Incubation: Around 80 days, hatching in April and early May.
Fledging: Similar to the other Wanderers, around 240 days, with young birds leaving in November and December.
Breeding success: In the region of 27 per cent, which is low in comparison to similar species, with 75 per cent of mortality occurring during the chick stage, linked to winter predation by the introduced house mouse.

Food
Squid appears to be the most important food, along with fish, probably crustaceans, and offal from fishing vessels.

Threats
Fishing: As with other *Diomedea*, the Tristan is a scavenger and known to be threatened by unregulated commercial fishing activities. Recent

calculations based on satellite tracking research estimate that up to 500 individuals may be killed each year, representing a significant proportion of the total population.

Predation: Uniquely, the breeding population on Gough Island is threatened by the introduced house mouse, *Mus musculus*. This predation starts once the chicks have become large enough to be left unguarded at four to five weeks. The mice, which are the largest known examples of their species, attack by chewing the rear part of the abdomen and thighs, several mice often working together. The chicks apparently have no defence mechanism and do not react significantly to the attacks. A 2007 feasibility study is proposing an eradication plan, but given the isolation of the island and its rugged and boggy terrain, matted vegetation and predictably cloudy self-generated weather, this will not be easy. Mice are more difficult to eradicate than rats, due to their smaller home ranges. This will be the largest such effort undertaken to date.

Amsterdam Albatross *Diomedea amsterdamensis*

Alternative or previous names: Wandering albatross. *Diomedea exulans*.
First described: Roux et al., 1983.
Taxonomic source: Brooke (2004), Robertson and Nunn (1998), Sibley and Monroe (1990, 1993).
Taxonomic note: Originally considered a subspecies of the wandering albatross, *D. exulans*, and while first described as a separate species by Roux et al. (1983) this split was not universally accepted until Robertson and Nunn (1998) provided supporting genetic evidence.
Origin of name: After breeding location, Amsterdam Island, Indian Ocean.
Conservation status: Critically Endangered.
Justification: A very small population of about 130 birds breeding in a confined area on one island. Additionally, it is threatened by two avian viruses which are affecting breeding success. Commercial fishing is also a potential threat. Endemic to Amsterdam Island.

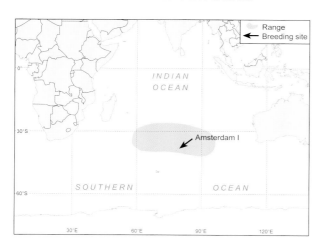

Description
The darkest of all the great albatrosses. While plumage gradually lightens with age, its whiteness never approaches even that of the Antipodean, let alone the Tristan or wandering albatross.
Identification: Dark cutting edge to upper mandible distinguishes it from other Wanderers at close range. Although not likely to overlap in range, it is most easily confused with female Antipodean albatross, though the Amsterdam is generally browner. Adults are easily confused with juveniles of any of the Wanderers.

Adult male
Body and rump: Largely mid-brown with exception of ventral area which is white with some vermiculated blotches adjacent, giving slightly mottled appearance.
Head and neck: Mid-brown apart from white face and chin, extending to throat in older birds. Eye dark.
Bill: Large and flesh-pink with pale horn-coloured hooked tip. Dark cutting edge to upper mandible, similar to royals.
Wings and tail: Upper wing black-brown becoming mottled brown close to mantle. Underwing white apart from dark tips and narrow dark trailing edge. Tail dark brown.

Adult female
Similar but generally browner overall, often with only white face and chin and white or light brown belly.

Juvenile
Almost entirely dark brown with white face and chin becoming gradually lighter with age. Bill dull grey-pink.

Size
Males significantly larger than females, except in weight.
Body length: Male 110 cm. Female 100 cm.
Wingspan: 3.00–3.25 m.
Weight: Male 6.0–8.0 kg. Female 5.0–7.0 kg.
Bill: Male 142–156 mm. Female 138–145 mm.
Egg: Length 117–125 mm. Diameter 73–77 mm. Weight 380–445 g.

Population and distribution
Total population is around 130 individuals of which some 80 are mature adults, with 18–25 pairs breeding annually. Monitoring since 1983 shows a population increase averaging seven per cent annually, from five eggs laid in 1984 to over 30 in 2001. This trend may now be overturned by the likely impact of recently spreading disease.

Oceanic range
Not well known due to small population and difficulty of identification at sea, foraging mainly within 1500 kilometres (932 miles) of breeding grounds while nesting, but extending 2200 kilometres (1367 miles) to the west of the island. Possibly ranges as far as Tasmania and New South Wales, in Australian waters, at other times.

Breeding
Only on Amsterdam Island in the Indian Ocean, nesting on the central plateau, at a height of around 500 metres (1640 ft). Birds return late January–February, with males arriving first. The whole breeding cycle is some two months later than the rest of the Wanderer superspecies. The nest is a typical low mound of vegetation held together with mud. Immatures first return at age four, more usually five, and may start breeding at seven or eight, but more usually 10 years of age.
Courtship: Very similar to the other great albatrosses, sky-pointing with bill uplifted, bill clacking, wing spreading, various vocalisations.
Laying: Single egg in latter half of February or early March.
Incubation: Around 80 days with hatching taking place in early to mid-May.
Fledging: Around 235 days with young leaving breeding grounds from mid-January through late February.
Breeding success: Around 70–75 per cent with successful pairs normally breeding biennially. Pairs produce on average one egg every 1.89 years and one fledgling every 2.4 years.

Food
Not well known due to small population, but likely to be largely squid and other cephalopods, with some fish, crustaceans and offal. As with other Wanderers, the Amsterdam is a scavenger.

Threats
Fishing: As a scavenger, the Amsterdam albatross is known to be at risk from commercial fishing along with the other Wanderers.
Predation: Rats and feral cats may prey on chicks; subantarctic skua can take unguarded eggs.
Habitat loss: Reduced breeding area due to introduced cattle and alteration of nesting grounds from past efforts at peat bog drainage.
Diseases: Although as yet unproven, avian cholera and *Erysipelothrix rhusiopa* viruses may be spreading from the nearby colony of Indian yellow-nosed albatross, and could be the explanation for high chick mortality in recent years. There is conjecture that these diseases were introduced by poultry brought to the island for human consumption.
Summary note: With such a very small population there is a very real prospect that the Amsterdam albatross could become effectively extinct within a few years, though individual birds may survive for several decades. The critical size for a viable breeding population is unknown. Unless immediate action is taken to address the threats from cattle, cats, fishing and avian diseases the Amsterdam albatross may be the first albatross to become extinct in historic times.

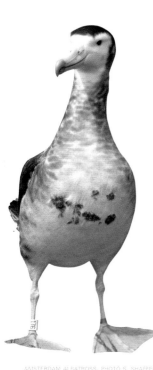
AMSTERDAM ALBATROSS, PHOTO S. SHAFFER

Northern Royal Albatross *Diomedea sanfordi*

Alternative or previous names: Royal albatross, *D. exulans mccormicki*, *D. epomophora*.
First described: Murphy, 1917.
Taxonomic source: Brooke (2004), Robertson and Nunn (1998).
Taxonomic note: *Diomedea epomophora* (Sibley and Monroe 1990, 1993) has been split into *D. epomophora* and *D. sanfordi* following Robertson and Nunn (1998) and Brooke 2004.
Origin of name: After Rollin Brewster-Sanford, sponsor of the Brewster-Sanford expedition during which the type specimen was shot by Rollo Beck in 1913. Interestingly, Murphy, who originally described the species, came to believe that he was mistaken and that the specimen was in fact *D. epomophora*, the southern royal; however, Robertson and Nunn (1998) vindicated Murphy's original description.
Conservation status: Endangered.
Justification: Breeding is almost entirely restricted to The Sisters and The Forty-Fours in the Chatham Islands, both of which are subject to occasional storms which can affect breeding success. The species is also known to be at risk from commercial fishing operations. Endemic to New Zealand.

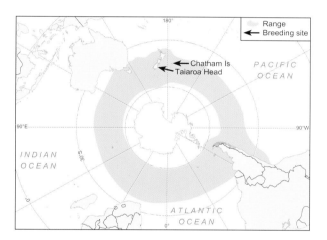

Description
The northern royal is a handsome and very white albatross with nearly all black upper wings, smaller but otherwise very similar physiologically to the southern royal, with which it does occasionally interbreed.
Identification: Distinguished from southern royals by all dark upper wings and from the Wanderers by this and all white body, mantle and tail. The bill is less pink than in Wanderers and the dark cutting edges of the mandibles, shared only with the southern royal and Amsterdam albatrosses, are also diagnostic.

Adult male and female
Sexes similar.
Body and rump: Entirely white including mantle.
Head and neck: White, though some females may show small amount of black on cap soon after moult. Eye dark.
Bill: Massive, flesh-pink, heavily hooked, with dull yellow tip. Black cutting edges to mandibles.
Wing and tail: Upper wing black. Underwing white apart from black tip extending to the carpal joint on leading edge, and a very thin black trailing edge.

Juvenile
Similar to adult but occasional dark flecking on back and tail; bill duller, almost grey.

Size
Males somewhat larger than females.
Body length: 115 cm.
Wingspan: Up to 3.25 m.
Weight: Up to 9 kg.
Bill: Male 165–172 mm. Female 154–160 mm.
Egg: Length 117–132 mm. Diameter 73–84 mm. Weight 378–465 g.

Population and distribution
Estimated annual breeding population of 6500–7000 pairs, 99 per cent of these in the Chatham Islands east of New Zealand, namely on The Forty-Fours and The Sisters. Also, this is the only 'mainland breeding' albatross, with a colony of some 30 pairs on Taiaroa Head in the South Island of New Zealand. Although increasing at this site, the overall population is thought to be declining.

Oceanic range
Circumpolar. Immatures and non-breeding adults found in all the southern oceans south of 40 degrees south, infrequently reaching Antarctic waters. Known to achieve complete circumnavigations between breeding years. Primary feeding areas located on the Patagonian Shelf as well as the Pacific side of South America.

Breeding
Nesting takes place primarily on plateaus on top of The Forty-Fours and The Sisters, often in vegetated areas. Nesting is denser than with any other great albatross due to lack of space. First breeding at anything from six to 22 years, but more normally at eight to nine years.
Courtship: Very similar to the other great albatrosses.

Laying: A single white egg with reddish blotches at the larger end, laid late October through late November.
Incubation: 77–80 days, hatching in mid-January through early February.
Fledging: Chicks brooded for 34 days, guarded for a further six. Fledging after 240 days, in September and October.
Breeding success: High, around 90 per cent of all eggs laid lead to fledging, with greatest mortality tending to occur after fledging. Random count on The Forty-Fours in February 2006 showed 90 per cent of nests had recently hatched chicks, with the balance still on eggs.

Food

Mostly squid and other cephalopods with some small fish, salps and occasionally crustaceans taken from the surface and up to a metre deep. Interestingly, bottom-dwelling octopus also figure in the diet of the colony on Taiaroa Head, presumably surface caught following spawning or from human or fur seal fishing activity.

Threats

Fishing: A known scavenger, it is at risk from commercial fishing, and while the New Zealand fisheries practise some mitigation measures, this is not the case with unregulated ships in international waters. Problem areas are known to exist in Chilean, Peruvian, Argentine and Brazilian fisheries.
Predation: Some human predation may still occur in the Chatham Islands, but not on a regular basis as in past tradition. However, given the small population and the threat from fishing, any such additional pressure is likely to endanger the species further. Skuas will attack unattended eggs or small chicks. Some chick mortality is possibly due to blowflies.
Habitat loss: In 1985 a large storm destroyed much of the vegetation on The Forty-Fours, thus reducing breeding activity and success for many years. However, by 2006 the vegetation had recovered and breeding appeared normal, with sitting birds almost totally concealed in dense cover provided by the Chatham Island button daisy *Leptinella featherstonii*.

NORTHERN ROYAL ALBATROSS, THE FORTY-FOURS, CHATHAM ISLANDS

Southern Royal Albatross *Diomedea epomophora*

Alternative or previous names: Royal albatross.
First described: Lesson, 1825.
Taxonomic source: Brooke (2004), Robertson and Nunn (1998).
Taxonomic note: *Diomedea epomophora* (Sibley and Monroe 1990, 1993) has been split into *D. epomophora* and *D. sanfordi* after Robertson and Nunn (1998) and Brooke (2004).
Origin of name: *Epomophora* from the Greek, refers to its slightly hump-backed appearance, having a rather bear-like posture of the neck and shoulders.
Conservation status: Vulnerable.
Justification: Very restricted breeding range consisting of four islands in the Auckland and Campbell Island groups south of New Zealand. Population is assumed to be stable, but is at risk from commercial fishing and possible habitat loss. Endemic to New Zealand.

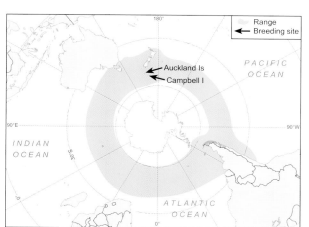

Description
The largest of all albatrosses, a magnificent, largely white bird, similar in most respects to the northern royal except for slight size difference and wing colour.
Identification: Distinguished from northern royal by the greater amount of white on upper wings, but less so than adult male wandering albatross. Also differs from mosty Wanderers by whiteness of head and tail, paler bill, and dark cutting edges to mandibles. It furthermore lacks the pink ear coverts frequently present in Wanderers.

Adult male and female
Sexes similar.
Body and rump: Almost entirely white, some dark vermiculated markings on mantle in younger adults.
Head and neck: White.
Bill: Very large, flesh coloured with dull yellow hooked tip. Dark line on cutting edges of mandibles.
Wings and tail: Upper wing black and white, with black primaries, secondaries and primary coverts, rest white or near white. Underwing white with black tip and very narrow black trailing edge.

Juvenile
Similar to adult, but more black on wings, with dark flecking on the mantle and a touch of black on the cap.

Size
Males slightly larger than females.
Body length: 107–122 cm.
Wingspan: 3.05–3.60 m.
Weight: Male 9.6–11.5 kg. Female 7.1–8.3 kg.
Bill: Male 180–190 mm. Female 166–178 mm.
Egg: Length 118–131 mm.

Population and distribution
Breeds primarily on Campbell Island, although a few hybridise with northern royals at Taiaroa Head on South Island, New Zealand. The population on Campbell is around 8400 annual breeding pairs. This constitutes 99 per cent of the overall population, with small numbers on Enderby and a further few pairs on Adams, and Auckland Islands. The population is considered stable, with some indication that it may be expanding.

Oceanic range
Very similar to northern royal. Circumpolar for immatures and non-breeding adults, though most

SOUTHERN ROYAL ALBATROSS

likely to be found from Great Australian Bight through to eastern South Atlantic, especially the Patagonian Shelf. Also ranges up the west coast of South America to Peru.

Breeding
Generally biennial, though some successful pairs may breed annually and 15 per cent of successful breeders and up to 24 per cent of failed breeders may skip a year and attempt breeding once every three years. Birds return late October and November, males before females. Nests are spread out among tussock grass on high ground, but avoiding windy ridges; low density in comparison to northern royal. Age at first breeding is eight to nine years, as with the northern royal.
Courtship: Very similar to all other great albatrosses, but generally less exuberant, with wing stretching and calling less frequent.
Laying: Single white egg with pinkish spots at larger end, laid late November through late December.
Incubation: 79 days on average, hatching in mid-February through mid-March.
Fledging: Brooding and guarding stage 35–42 days. Fully fledged after 225–250 days, in October and November.
Breeding success: 50–60 per cent but varies annually.

Food
Largely cephalopods such as squid, with some fish, crustaceans and salps. Surface feeder, but may make shallow dives to two metres.

Threats
Fishing: As a scavenger, is at risk from commercial fishing operations, particularly longliners, although very few recorded kills in New Zealand waters, as it is less bold than others around fishing operations.
Predation: Apparently a rare occurence, a rampaging bull sea lion killed many incubating adults high in the Mount Lyall area on Campbell Island.
Habitat loss: On Campbell and Enderby there is a danger of loss of habitat due to the spread of *Dracophyllum* and other shrubs after the removal of grazing mammals.
Other: On Campbell, anecdotal evidence of several incubating adults appearing to have been killed by large hailstones during a fierce summer squall around Mount Lyall. Over 35,000 birds were banded between 1941 and 1998, mainly by untrained Met Service staff stationed on Campbell, resulting in high levels of injuires over time (7 per cent in those banded as chicks). To remedy the problem, DOC has removed or replaced all bands in a four-year programme.

LEG DAMAGE CAUSED BY HISTORIC BANDING BY UNTRAINED MET SERVICE STAFF ON CAMPBELL ISLAND.

Black-browed Albatross *Thalassarche melanophrys*

Alternative or previous names: *Diomedea melanophrys*.
First described: Temminck, 1828.
Taxonomic source: Brook (2004); Robertson and Nunn (1998).
Taxonomic note: *Diomedea melanophrys* (Sibley and Monroe 1990, 1993) has been split into *melanophrys* and *impavida* and both transferred to the genus *Thalassarche* following Robertson and Nunn (1998) and Brooke (2004).
Origin of name: Both common and scientific names refer to its black eye marking.
Conservation status: Endangered.
Justification: While still one of the most numerous of all albatrosses, it has, however, suffered very significant population declines, especially in the Falkland Islands, its main breeding grounds. Some recent information indicates that this decline may have lessened or ceased. It is one of the species most affected by commercial fishing.

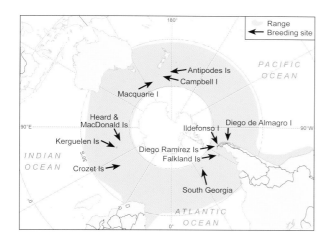

Description
A striking, large black and white seabird with a white head, a bright yellow-orange bill and very distinctive black shadowing around the eye.
Identification: Distinguished from Campbell albatross by dark eye, smaller but more intense black 'brow' and by greater amount of white on underwing, and from Chatham by white head. Juveniles indistinguishable from Campbell but from juvenile grey-headed albatrosses by yellow on bill and less grey on head and neck.

Adult male and female
Sexes similar.
Body and rump: All white, also undertail coverts. Mantle black or near black.
Head and neck: All white but with distinctive triangular black shading around and forward of the eye. Eye dark, with white crescent behind it. Distinctive pink gape line visible only when displaying or feeding young.
Bill: Bright peach-orange with distinctly hooked tip.
Wings and tail: Upper wing black with slight greying on mantle; underwing white but with black tip, narrow black trailing edge and broad black area along the leading edge. Tail black.

Juvenile
Similar to adult but has pale grey head and especially neck, sometimes extending to a full collar. Bill dull yellow with a distinctive black tip. Underside of wing generally grey with darker tip and edges.

Size
Males generally larger than females.
Body length: 80–95 cm.
Wingspan: 2.10–2.50 m.
Weight: Male 3.3–4.7 kg. Female 2.8–3.8 kg.
Bill: Male 116–122 mm. Female 114–121 mm.
Egg: Mean length 104 mm. Diameter 66 mm. Weight 260 g.

Population and distribution
Most widespread and numerous of all albatrosses, but with vast majority breeding in the Atlantic. Total population is estimated at 530,000 pairs, of which 60 per cent nest in the Falkland Islands (mainly Grand and Steeple Jason, and Beauchêne), 20 per cent on South Georgia (Willis, Annekov, Cooper and Bird Islands), and 20 per cent in Chile (Diego Ramirez, Ildefonso and Diego de Almagro). The population has declined significantly — by at least 24 per cent — since the mid-1990s when Brooke (2004) estimated 702,000 breeding pairs based on data up to 2000. While anecdotal evidence (Strange, unpublished pers. com. 2007) now suggests that the decline is no longer continuing in some Falkland colonies, the threat remains, especially in view of the

species' propensity to follow fishing vessels. In the New Zealand region, small numbers (c. 150 pairs) at Bollons Island (Antipodes), c. 25 pairs mixed with Campbell albatrosses at Campbell Island, and a few birds (one breeding pair at least) at The Snares Western Chain.

Oceanic range

Circumpolar with breeding populations in three southern oceans, though primarily concentrated in the Atlantic. Birds seen in vicinity of New Zealand and Australia more likely to be Campbell albatross, from which it can only be distinguished at close range. Occasionally ventures north of the equator, with a few records as far north as Scotland.

Breeding

Generally, adults arrive at their breeding islands in late September or early October and young leave in April. Immatures return as young as two years old, but do not start breeding until eight to 12 years. Usually nests on grass and tussock-clad slopes and ledges; however, in larger colonies on flat ground there is little or no vegetation within the colony itself. Inland the birds create 'runways' for taking off and landing. The nest is a typical truncated cone with deep bowl, mainly constructed of mud bound with vegetation, feathers and anything else they can find. Nests are repaired each year and used repeatedly. An annual breeder, but a significant proportion of both successful (25 per cent) and failed (33 per cent) breeders do not breed the following year.
Courtship: Similar to other mollymawks, but closely packed nests make any courtship ritual modest; includes the usual assortment of billing, preening, tail fanning and bowing with loud vocalisations.
Laying: A single white egg with some spotting, laid in late September and early October.
Incubation: Averages 70 days, hatching mainly late December.
Fledging: 116–125 days but varies with location, birds leave the colony in April.
Breeding success: Quite variable, with tick *Ixodes uriae* infestation resulting in significant chick mortality in some colonies. In South Georgia survival is around 27 per cent, with 62 per cent of eggs hatching and 43 per cent of chicks fledging. Breeding success also linked to availability of food.

Food

Varies according to population. In South Georgia most important food is krill, *Euphausia superba*, which can constitute up to 40 per cent of diet, followed by squid and small fish, the latter being the primary food source in the Falklands' population. Opportunistic diet varies with availability, obtained by lunging within a metre of the ocean surface. Frequently scavenges around fishing vessels.

Threats

Fishing: One of the most gregarious of all albatrosses, black-browed are particularly at risk from pelagic fisheries. Most of the decline since the mid-1990s is attributed to fishing, particularly longlining.
Predation: Immatures at risk from pinnipeds, sharks and southern giant petrels, the petrels taking large chicks from the nest in some colonies, e.g. Beauchêne. Also some loss of eggs and chicks to subantarctic skua and striated caracara in the Falklands and Chilean islands. Anecdotal evidence shows that they are sometimes deliberately targeted as food by pelagic fishermen in the South Atlantic.
Disease: Some chick loss due to tick infestation at some nesting sites.

Campbell Albatross *Thalassarche impavida*

Alternative or previous names: Black-browed albatross, *Diomedea melanophrys*, *T. melanophrys*.
First described: Matthews, 1912.
Taxonomic source: Brooke (2004); Robertson and Nunn (1998).
Taxonomic note: *Diomedea melanophrys* (Sibley and Monroe 1990, 1993) has been split into *melanophrys* and *impavida* and both transferred to the genus *Thalassarche* following Robertson and Nunn (1998) and Brooke (2004).
Origin of name: Common name after breeding island. *Impavida* is Latin for 'fearless', possibly in reference to its pale eyes imparting a fearless stare. Name first used by Swedish botanist Daniel Solander to describe a specimen he shot on Cook's first voyage in 1770.
Conservation status: Vulnerable.
Justification: Breeding is restricted to a single location, the northern end of Campbell Island in the New Zealand Subantarctic. Population declined markedly from 1970 to the early 1980s, followed by a slight increase 1984–97. Known to be at risk from commercial fishing. Endemic to New Zealand.

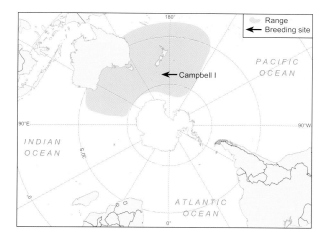

Description
A medium-sized black and white albatross with a dramatic yellow-orange bill and black eye shadow. Very similar to the black-browed albatross, with which it was previously identified as the same species.
Identification: Distinguished from black-browed by pale iris, larger patch of eye shadow, which extends further forward, and less extensive white on underwing. From shy and white-capped by bright yellow bill and from Chatham by white head. Juvenile is indistinguishable from juvenile black-browed, but differs from juvenile grey-headed by lighter-coloured bill with dark tip, and lighter grey on head and neck.

Adult male and female
Sexes similar.
Body and rump: All white apart from mantle which is dark grey or black.
Head and neck: All white apart from black triangular area around eye. Bright pink gape line, visible only when displaying or feeding chick. Pale greyish or straw-coloured iris is distinctive and diagnostic; white crescent behind eye.
Bill: Bright orange-yellow becoming peach at tip.
Wings and tail: Upper wing black with slight greying on mantle. Underwing white with black edging, deeper on the leading edge. Tail black.

Juvenile
Similar but underwing darker, generally grey rather than white. Grey nape and neck, sometimes extending to full collar. Bill dull yellow with dark tip.

Size
Males are generally larger than females.
Body length: 80–95 cm.
Wingspan: 2.1–2.5 m.
Weight: Male 2.75–3.8 kg. Female 2.2–3.2 kg.
Bill: Male 105–118 mm. Female 105–114 mm.
Egg: Length 102 mm. Diameter 66 mm. Weight c. 250 g.

Population and distribution
Thought to number around 23,500 annual breeding pairs (1995–97), with colonies located at the northern end of Campbell Island and its satellite islet Jeanette Marie. Previous sharp population decline appears to have stabilised and increased slightly 1984–97.

Oceanic range
Ranges largely to the south of New Zealand and Australia, to the edge of the Ross Sea, also in the Tasman Sea and South Pacific to the east and north of New Zealand.

Breeding
Builds a typical pedestal nest of mud and vegetation some 30 centimetres (1 ft) in diameter and up to 45 centimetres (1.5 ft) high, densely spaced on grass and tussock-covered ledges and slopes, often creating quagmires from which the nests emerge as islands. Pairs use the same nest in successive seasons, resulting in it gradually growing in size. Average age at first breeding is 10 years, earliest six years.
Courtship: Similar to other mollymawks, especially

CAMPBELL ALBATROSS, CAMPBELL ISLAND

black-browed, with which it occasionally interbreeds. A combination of nodding, bowing, bill fencing, tail fanning, and gaping accompanied by typical loud vocalisations.

Laying: A single white, occasionally pinkish-white, egg with variable ring of small reddish spots at blunt end, laid late September through mid-October.

Incubation: 65–72 days, with hatching in early December.

Fledging: Brooding lasts 16–26 days, with fledging at 130 days in April and May.

Breeding success: Around 66 per cent, with 86 per cent of eggs hatching and 79 per cent of chicks surviving to fledge. Survival rate to age five, when immatures first return to breeding area, is 28 per cent, indicating that less than 22 per cent of hatched chicks become breeders. Annual survival rate of breeding adults is, or was (1984–95), around 95 per cent.

Food
Variable, with breeding birds feeding on the Campbell Plateau, primarily on southern blue whiting, while non-breeders feed in pelagic waters taking mainly squid and other cephalopods. Largely surface feeding, but also scavenges around fishing vessels.

Threats
Fishing: A gregarious scavenger, making it particularly vulnerable to industrial fishing methods.
Predation: Eggs and chicks liable to predation by subantarctic skua.

White-capped Albatross *Thalassarche steadi*

IMMATURE WHITE-CAPPED ALBATROSS

NOTE Most previously published range maps for this species indicate a circumpolar distribution. This was caused by confusion deriving from before the taxonomic split of this species from the Chatham and Salvin's albatrosses.

Alternative or previous names: Auckland shy albatross, *Diomedea cauta steadi*.
First described: Falla, 1933.
Taxonomic source: Robertson and Nunn (1998).
Taxonomic note: *Diomedea cauta* (Sibley and Monroe 1990, 1993) has been split into *cauta*, *eremita* and *salvini* and all transferred to the genus *Thalassarche* following Brooke (2004). *T. cauta* has subsequently been split into *cauta* and *steadi* following Robertson and Nunn (1998). The name *steadi* was first proposed by Falla (1933).
Origin of name: Common name descriptive; Latin after David George Stead (1877–1957), Australian naturalist.
Conservation status: Near Threatened.
Justification: Very concentrated breeding sites with about 95 per cent of the population on Disappointment Island in the Auckland group south of New Zealand. It is known to suffer significant losses in commercial fisheries off South Africa, and so it is possible that the status will be reviewed in the foreseeable future. Endemic to New Zealand.

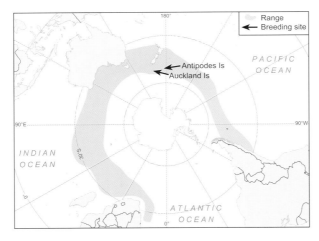

Description
The largest of the mollymawks together with the shy albatross of Tasmania, from which it has recently been taxonomically separated on the basis of DNA analysis. Typical black and white plumage as with others of the genus, but with a noticeably white cap and forehead.

Identification: Virtually indistinguishable from shy, though top ridge on bill slightly more yellowish, especially near base, whereas in shy bill is more uniformly coloured. Distinguished from Salvin's by whiter cap and much lighter cheek and neck, plus greenish-yellow bill tip, and smaller (or absence of) dark patch at tip of lower mandible. The head and neck, and horn-coloured bill with greenish-yellow tip, are overall paler than related species. Juvenile shy and white-capped indistinguishable; also easily confused with juvenile Salvin's and Chatham though possibly the White-capped's paler grey hood and grey collar may help differentiate.

Adult male and female
Sexes similar.

206 • Species Profiles

Body and rump: White but with dark or black mantle.
Head and neck: Pale grey, especially around nape and cheeks, with white cap and distinctive dark triangular eye patch. Bright orange gape line, visible only when displaying and feeding young.
Bill: Dull yellowish-grey with greenish-yellow tip. Bright orange skin at base of lower mandible. May also have dark patch on tip of lower mandible, especially in younger birds.
Wings and tail: Upper wing black with slightly greyer mantle. Underwing white but with narrow dark leading and trailing edges, with tip also dark. Tail dark.

Juvenile
Very similar to adult but head and nape grey, extending to form a narrow collar. Bill more grey and with a dark tip.

Size
Males generally larger than females.
Body length: 90 cm.
Wingspan: 2.12–2.56 m.
Weight: Male 3.3–5.3 kg. Female 2.6–4.2 kg.
Bill: Male 136–141 mm. Female 126–139 mm.
Egg: Length 95–112 mm. Diameter 60–94 mm.

Population and distribution
Total population thought to be about 110,000 annual breeding pairs, with about 95 per cent nesting on Disappointment in the Auckland group, and the remainder on Adams and Auckland itself, plus 20 pairs on Bollons Island near Antipodes. There is some indication that numbers may be increasing but more data is needed for a detailed profile.

Oceanic range
Non-breeders and immatures leave their breeding islands and generally travel west across the Indian Ocean, with considerable concentrations found off southern Africa and the coast of Namibia, where high mortality rates have been noted. Earlier reports of movement towards South America appear to pertain to Salvin's albatross (previously considered the same species) but are not supported by satellite tracking. Nesting birds remain mostly in New Zealand waters, with occasional forays to Tasmania.

Breeding
An annual breeder building typical mollymawk truncated cone and bowl. Birds return to their colonies in late October, but there remains a dearth of information on its breeding cycle which is only just being addressed by new research.
Courtship: Similar to all mollymawks, but rather less loud, with tail fanning, bill touching, gaping and fake preening the most prominent gestures.
Laying: Single white egg with reddish brown speckles; inferred start of laying in late November.
Incubation: Likely to be 65–70 days in line with other mollymawks, with hatching inferred to start in early February.
Fledging: Similar to other mollymawks, around 140 days, fledging inferred from early June.
Breeding success: No data available.

Food
Mainly cephalopods and fish, particularly mackerel and redbait, caught on the surface or by pursuit plunging using wings for propulsion, staying submerged for up to five seconds. Also, aggressively scavenge for offal from fishing vessels.

Threats
Fishing: It is estimated that up to 8000 individuals may be killed each year in both trawling and longline fisheries off southern Africa. Current BirdLife programmes are helping to reduce this. Mortality in New Zealand waters is being curbed with better use of mitigation techniques, but it is still one of the most heavily impacted species.
Predation: On Auckland Island nesting is affected in areas accessible to feral pigs. Predation by subantarctic skuas on young chicks appears to be heavy around the periphery of some colonies. Immature albatrosses are potentially at risk from sharks and giant petrels, but no observations are available.

Shy Albatross *Thalassarche cauta*

Alternative or previous names: Tasmanian shy albatross; *Diomedea cauta*.
First described: Gould 1841.
Taxonomic source: Brooke, 2004.
Taxonomic note: Previously *Diomedea cauta* (Sibley and Monroe 1990, 1993) but split into *cauta* (shy), *eremita* (Chatham) and *salvini* (Salvin's) and transferred to genus *Thalassarche* following Brooke (2004). Subsequently, *T. cauta* split into *cauta* and *steadi* after Robertson and Nunn (1998). The name *steadi* was first proposed by Falla (1933).
Origin of name: From the Latin cauta which translates as 'cautious', reflecting the English name 'shy', yet there seems no obvious reason for this name, as behaviourally *T. cauta* is similar to the other mollymawks. A possible alternative explanation is that it has a 'shy' or 'modest' look.
Conservation status: Near Threatened.
Justification: The species has a very restricted breeding range around Tasmania, and while numbers are stable or possibly increasing, the total breeding population is still small. There is additionally a threat to breeding success from avian pox. Endemic to Australia.

NOTE Most previously published range maps for this species indicate a circumpolar distribution. This was caused by confusion deriving from before the taxonomic split of this species from the Chatham and Salvin's albatrosses.

SHY ALBATROSS, ALBATROSS ISLAND, AUSTRALIA. PHOTO B. BAKER

Description
A medium-sized black and white albatross with a distinctive white cap, dark eye shadow and dull yellowish-grey bill nearly identical to the white-capped albatross with which it was formerly 'joined'.
Identification: Virtually indistinguishable at sea from white-capped though latter has slightly more yellowish upper ridge to bill. Differs from Salvin's by generally lighter neck and cap and greenish-yellow bill tip, plus smaller, or absence of, dark patch at tip of lower mandible. Juvenile shy and white-capped indistinguishable; also easily confused with Salvin's and Chatham, though possibly paler grey hood and grey collar.

Adult male and female
Sexes similar.
Body and rump: White but mantle dark grey to black.
Head and neck: Pale grey but with distinctive white cap. Dark triangular eye patch, eye dark with white crescent behind it. Bright orange gape line visible only during display and chick feeding.
Bill: Dull yellowish-grey with distinctive bright greenish-yellow tip and some variable yellow colouring on culmen (upper ridge), slightly brighter at base. Bright orange skin at base of lower mandible, plus dark patch at tip in younger birds.
Wings and tail: Upper wing black but with a rather greyer mantle. Underwing white with narrow black leading and trailing edges, wingtips also dark but less extensive than others in same genus. Tail black.

Juvenile
Very similar to adult, but head greyer and white cap less obvious; grey collar diagnostic. Bill greyer, with black tip.

Size
Males generally slightly larger than females.
Body length: Male 90–110 cm.
Wingspan: 2.12–2.56 m.
Weight: Male 3.9–5.1 kg. Female 3.2–4.4 kg.
Bill: Male 128–138 mm. Female 122–132 mm.
Egg: Length 95–121 mm. Diameter 60–94 mm.

Population and distribution
Driven nearly to extinction by the feather trade in the late nineteenth century, it was reduced to fewer than 300 pairs by 1909, but has recovered strongly and numbers may still be increasing. Three distinct breeding populations, totalling around 12,200 annual breeding pairs, all in the vicinity of Tasmania: Albatross Island in Bass Strait, Mewstone and Pedra

Branca with a small number (2 per cent of total), both off southern Tasmania.

Oceanic range
Little reliable data due to recent splitting of species into shy and white-capped, and previously from Salvin's and Chatham. However, recent satellite telemetry studies show that adults remain mostly within Australian waters between 30 and 50 degrees south, while juveniles travel across the Indian Ocean to waters off South Africa and Namibia.

Breeding
Nest is a typical mollymawk truncated cone and bowl of mud and debris. Adults return to breeding area in early September, males generally somewhat ahead of females. Immatures may return to colony after two to three years and begin breeding at age five to six, which is earlier than most others in this genus.
Courtship: Similar to all other mollymawks, especially white-capped.
Laying: Single egg laid mid–late September. White with reddish brown spots at larger end.
Incubation: Averages 70 days, hatching in December.
Fledging: Brood stage is 20–25 days with fledging after 120–140 days, generally in April.
Breeding success: On Albatross Island fledging success rate varies between 20 per cent and 50 per cent.

Food
Forages largely by surface lunging, though has been recorded reaching at a depth of 7.4 metres (24 ft), but with most dives less than 3.5 metres (11.5 ft). Feeds on small fish including mackerel, squid and other cephalopods, crustaceans and tunicates, plus offal from fishing vessels.

Threats
Fishing: Habitual scavenger around fishing vessels, and a frequent casualty from tuna longline operations in Australian waters, but reduced since this fishery has moved out of the area. For juveniles migrating across the Indian Ocean to the southern Africa region, both longline and trawl fisheries remain serious threats.
Predation: No significant predation. Immatures potentially at risk from sharks and giant petrels, but no observations available.
Habitat loss: Pedra Branca is subject to occasional extreme wave action which may affect breeding success. There is also some indication that the expansion of the Australasian gannet colony may be reducing the area available for breeding.
Disease: Avian poxvirus has been recorded in some birds on Albatross Island. This is a potentially serious threat though at present there is no indication of the vector, or of its impact.

SHY ALBATROSS, ALBATROSS ISLAND, AUSTRALIA. PHOTO B. BAKER

Salvin's Albatross *Thalassarche salvini*

Alternative or previous names: Shy albatross; *Diomedea cauta salvini*; *Thalassarche cauta salvini*.
First described: Rothschild, 1893.
Taxonomic source: Brooke (2004), Robertson and Nunn (1998).
Taxonomic note: *Diomedea cauta* (Sibley and Monroe 1990, 1993) has been split into *cauta*, *eremita* and *salvini* and all transferred to the genus *Thalassarche*, following Brooke (2004). *T. cauta* subsequently split into *cauta* and *steadi* following Robertson and Nunn (1998).
Origin of name: After Osbert Salvin (1835–98), British ornithologist.
Conservation status: Vulnerable.
Justification: While there is some concern about a decline in numbers, the chief justification is that 97 per cent of the population breeds at just one location, the Bounty Islands, an isolated group of rocks south-east of New Zealand. It is also known to scavenge and therefore is at risk from commercial fishing operations. Endemic to New Zealand.

SALVIN'S ALBATROSS

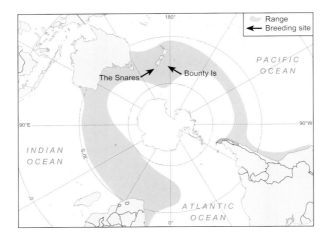

Description
A medium-sized black and white albatross, with a grey head and dull yellow and grey bill, making it rather less handsome than other mollymawks.
Identification: Distinguished from lighter shy and white-capped by ash-grey sides of neck and pale yellow stripe on upper bill ridge contrasting with grey sides, plus dark patches on lower tip. Dark underwing tips also more extensive. Chatham differs by much darker grey head and bright orange-yellow bill. Buller's and grey-headed have distinctive bright yellow and black bills, also more extensive black leading edge to underwings.

Adult male and female
Sexes similar.
Body and rump: White with white rump.
Head and neck: Uniform light grey blending into mantle, with typical mollymawk dark triangular eye patch. Bright orange gape line, which is only exposed during display and when feeding chick; presence indicated by a slight depression in cheek feathers.
Bill: Dull horn-coloured, with greyish sides and pale cream stripe along top ridge, yellowing near base; bill tip has a dark smudge on either side of lower mandibles. Brilliant orange skin at base of lower mandible.
Wings and tail: Dark upper wing becoming greyish on mantle and blending in to grey head and neck. Underwing largely white with narrow black leading and trailing edges, distinctive black thumb mark at base of leading edge of underwing, and black wingtip. Tail dark.

Juvenile
Similar to adult, bill darker with more distinctive black tip.

Size
Males somewhat larger than females.
Body length: 90–100 cm.
Wingspan: 2.10–2.50 m.
Weight: Male 3.1–4.3 kg. Female 2.8–4.2 kg.
Bill: Male: 113–121 mm. Female: 209–118 mm.
Egg: Length 102–110 mm. Diameter 66–70 mm.

Population and distribution
The main population, consisting of some 30,000 annual breeding pairs on Bounty Islands, with a further 1200 pairs on Western Chain Islets in The Snares group, both locations consisting of barren rocks. A few pairs have occasionally been recorded nesting on Chathams and Crozet Islands.

Oceanic range
Non-breeders and immatures range across Indian and Pacific Oceans to waters off South Africa and South America, travelling as far north as Peru.

Breeding
The bulk of the population breeds on the Bounty Islands which are devoid of vegetation for use in nest building. Instead of the typical mollymawk truncated cone, many nests consist of no more than a minimal crescent pasted to the downhill side of the slope, using bones, feathers, rock chips and fur seal hair, with guano as binding agent. Scarcity of nest-building material appears to be the limiting factor in habitat use. Birds return to nesting area in August. No information as to age of first breeding.
Courtship: A typical mollymawk affair, though limited by shortage of space and close proximity of surrounding nests, with very loud vocalisations.
Laying: Single white egg with some dark red spotting at larger end, laid in August–September.
Incubation: Around 72 days with most chicks hatching in early November.
Fledging: 140–150 days with fledging taking place in late March or April.
Breeding success: No data available, though population thought to be stable.

SALVIN'S ALBATROSS, BOUNTY ISLANDS

Food
No specific data available, but likely to be similar to Chatham, shy and white-capped albatrosses, mainly cephalopods and fish.

Threats
Fishing: A scavenger and therefore always at risk from commercial fishing operations; occurs in bycatch, particularly in demersal longline fisheries.
Predation: Giant petrels may take larger young and immatures.
Other: Some of the main nesting sites on Bounty can be washed over by large seas, although this does not usually occur during breeding season. Burgeoning New Zealand fur seal population is placing pressure on available nesting space.

Chatham Albatross *Thalassarche eremita*

Alternative or previous names: Chatham shy albatross, *Diomedea cauta eremita*. *Thalassarche cauta eremita*.
First described: Murphy, 1930.
Taxonomic source: Brooke (2004), Robertson and Nunn (1998).
Taxonomic note: *Diomedea cauta* (Sibley and Monroe 1990, 1993) has been split into *cauta*, *eremita* and *salvini* and all transferred to the genus *Thalassarche* following Brooke (2004). *T. cauta* subsequently split into *cauta* and *steadi* following Robertson and Nunn (1998).
Origin of name: Common name after breeding islands; Latin for 'hermit', a reference to its very restricted and isolated breeding location.
Conservation status: Critically Endangered.
Justification: Breeds on a single isolated volcanic stack, The Pyramid, in the Chatham Islands east of New Zealand. Total available breeding area is less than one hectare, part of which is liable to be washed over by large waves in extreme weather conditions. Also the species is known to suffer mortality from commercial fishing operations and additionally may still be the subject of traditional human predation. Endemic to New Zealand.

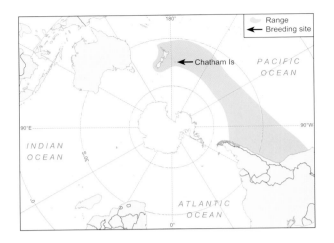

Description
One of the most attractive albatrosses, this is a typical mollymawk, but darker, with a distinctive grey head and handsome bright yellow-orange bill.
Identification: Distinguished from Salvin's, white-capped and shy by darker head and bright yellow bill; from Campbell and black-browed by grey head and more white on underside of wings.

Adult male and female
Sexes similar.
Body and rump: White with black or very dark mantle.
Head and neck: Dark grey with distinct triangular dark eye shading giving it a rather stern look, highlighted by white crescent behind the eye. The head and neck are darker than any other mollymawk. Bright orange gape line is only visible during display and chick feeding.
Bill: A striking bright chrome-yellow with a conspicuous dark tip to the lower mandible. Orange skin line at base of lower mandible.
Wings and tail: Upper wing near black, with somewhat greyer mantle. Underwing white with narrow black edge, wider at tip. Tail dark.

Juvenile
Similar to adults but generally greyer, dark horn-coloured bill with black tip which extends to both mandibles.

Size
Males are larger than females but the difference is less than with any others of the tribe.
Body length: 90–100 cm.
Wingspan: 2.10–2.56 m.
Weight: Male 3.6–4.0 kg. Female 3.1–3.9 kg.
Bill: Male 116–130 mm. Female 113–124 mm.
Egg: Length 98–106 mm. Diameter 65–69 mm.

Population and distribution
With a single population of around 5300 annual breeding pairs nesting on The Pyramid in the Chatham Islands, this species has the smallest breeding area of any albatross, and probably of any seabird. The actual area occupied is no more than one hectare of usable habitat. A few occur on The Snares Western Chain, but none has been associated with a chick.

Oceanic range
Breeding adults forage largely over the Chatham Rise and to the south of the Chatham Islands. Juveniles and non-breeders travel east to Tasmania and west across the Pacific to South America and up the coast to Peru, a favourite feeding area, returning via a more northerly route. Has recently been recorded off South Africa.

Breeding

The highest nesting density is found on ledges on the upper parts of The Pyramid, a precipitous sea stack to the south of the Chatham Islands. The nest is a typical mollymawk truncated cone and bowl, laboriously pasted to clumps of vegetation and rock slabs. However, in one area sheltered from rain under a large overhang, the nests attain substantial size and age as they are added to every year. Birds return to the colony during August. While juveniles may return at age four, the record of first breeding is seven years.

Courtship: Very similar to the other mollymawks, with the notable addition of short flights with showy landings accompanied by gaping and loud wails.

Laying: A single white egg laid in late August or early September, 90 per cent complete by 18 September.

Incubation: 66–72 days, with hatching in late October through end of November.

Fledging: Some 120 days, generally between late February and mid-April.

Breeding success: 50–65 per cent in normal years, but may fall to 35 per cent in years where hot or stormy weather has adverse effects.

Food

Very little detailed information but likely to be similar to other mollymawks, e.g. fish, squid and other cephalopods and some crustaceans.

Threats

Fishing: A scavenger and always at risk from commercial fishing operations, though not one of the species most seriously affected.

Predation: Large chicks are still harvested irregularly by the Chatham Islanders. There is a small amount of predation by subantarctic skua on chicks and unattended eggs.

Habitat loss: Vulnerable breeding area, the lower parts liable to be washed over by large waves. Recent trend toward hotter, drier weather, possibly reflecting global warming, has exacerbated loss of vegetation as nesting material.

Grey-headed Albatross *Thalassarche chrysostoma*

Alternative or previous names: Flat-billed, Gould's, grey-mantled albatross. *Diomedea chrysostoma, T. culminata.*
First described: Forster, 1785.
Taxonomic source: Brooke (2004), Robertson and Nunn (1998), Sibley and Monroe (1990, 1993).
Origin of name: From the Greek *krysos,* 'gold', and *stoma,* 'mouth', after its bill markings. Common name descriptive.
Conservation status: Vulnerable.
Justification: While this species breeds at sites in all three southern oceans there has been an apparent loss of 48 per cent of the breeding population over the last 90 years. The Bird Island (South Georgia) population has declined by 1.4 per cent per annum over a 20-year period, most likely linked to commercial fishing operations.

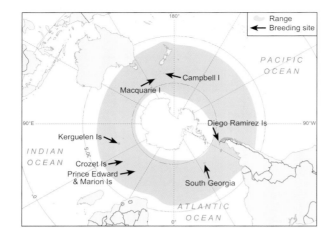

Description
A small black and white albatross with a grey head and distinctive bill markings, encountered throughout the southern oceans.

Identification: Bill with bright orange yellow upper and lower ridges shared with Buller's and to lesser degree with both yellow-nosed (upper mandible only); however, lacks pale cap and shows less grey on sides of neck than Buller's. Also Buller's has more extensive white on underwing. Salvin's and Chatham are larger, have very different bills, and also whiter underwing. Juvenile black-browed has less distinctive grey head and lighter bill with black tip.

Adult male and femals
Sexes identical.
Body and rump: White with dark grey or black mantle.
Head and neck: Mid-grey. Dark brown iris, with triangular dark eye shading and white crescent. Bright yellow gape line on cheek visible only during display and chick feeding.
Bill: Black with bright yellow-orange upper and lower ridges. Bright yellow skin at base of lower mandible.
Wings and tail: Upper wing black with slight greying on the mantle. Underwing white but with broad dark leading and trailing edges. Tail near-black.

Juvenile
Grey on head and neck less distinct but more extensive. Underwing largely grey with darker leading and trailing edges. Bill black.

Size
Males are slightly larger than females.
Body length: 70–85 cm.
Wingspan: 1.80–2.05 m.
Weight: Male 3.1–4.3 kg. Female 2.8–4.2 kg.
Bill: Male 113–121 mm. Female 109–118 mm.
Egg: Length 102–110 mm. Diameter 65–70 mm. Weight 262–290 g.

Population and distribution
This is one of the more widely distributed of all albatross species, with seven populations in three distinct groups: South Atlantic — South Georgia (Bird, Willis and other islands), Chile (Diego Ramirez); Indian Ocean — Marion and Prince Edward, Kerguelen, Crozet; South Pacific — Campbell and Macquarie. Total population estimated at 92,300 annual breeding pairs, with the largest contingent (60 per cent) at South Georgia.

Oceanic range
Circumpolar distribution; normally ranges between 46 and 64 degrees south in summer, and between 39 and 51 degrees south in winter, although occasionally reaching 15 degrees south off Peru. Breeding birds feed noticeably further afield than the black-browed (with which they often share colonies), travelling up to 12,000 kilometres (7500 miles) on a single trip during the incubation and

GREY-HEADED ALBATROSS, CAMPBELL ISLAND

GREY-HEADED ALBATROSS, CAMPBELL ISLAND

brooding period. During their non-breeding year, birds may circumnavigate the globe, sometimes more than once.

Breeding

Nests colonially, building a typical truncated cone and bowl, often on the periphery of other mollymawk colonies on cliff tops or among tussock grass. Adults return to breeding grounds in September and early October, but there is substantial variation between locations. Immatures may return at age three, more commonly six to seven, with first breeding at age eight, the average being 13.5 years, making this the latest breeding of all albatrosses. The only biennially breeding mollymawk, but there is no obvious reason for this. One suggestion — though with little supportive evidence — is that its main food contains less energy than that of other species such as the black-browed, which feeds largely on krill. A significant proportion of both failed and successful breeders defer breeding a further year.

Courtship: Very similar to all mollymawks, but more sedate and with mutual preening more prominent.
Laying: A single white egg with reddish brown spots laid during October.
Incubation: 69–78 days, hatching in December and early January.
Fledging: Average of around 140 days to fledging, with birds leaving colony in late April or May.
Breeding success: 40 per cent in South Georgia.

Food

A surface feeder, its main food is fish, squid and krill, in contrast to black-browed which takes more krill than squid. This may be an indicator of grey-headed being a more pelagic feeder while black-browed feeds on or near the continental shelf.

Threats

Fishing: As a surface scavenger, it is known to be at risk from commercial fishing operations, with most of the documented population decline thought to be due to this cause.
Predation: Subantarctic skua and snowy sheathbill may prey on chicks and eggs.

Buller's Albatross *Thalassarche bulleri*

Alternative or previous names: *Diomedea bulleri. Thalassarche platei.* Northern Buller's or Pacific albatross, *T. b. platei.* Southern Buller's albatross, *T. b. bulleri.*
First described: Rothschild, 1893.
Taxonomic source: Brooke (2004), Christidis and Boles (1994), Sibley and Monroe (1990, 1993), Stotz et al. (1996), Turbott (1990).
Taxonomic note: Retained as a single species (Sibley and Monroe 1990, 1993) following Brooke (2004) contra Robertson and Nunn (1998) who split it into two species, *T. bulleri* and *T. platei.* Currently accepted as two subspecies: *T. b. bulleri* and *T. b. platei.*
Origin of name: After Sir Walter Lawry Buller (1838–1906) author of *Birds of New Zealand.*
Conservation status: Near Threatened.
Justification: Small breeding population with only three significant breeding locations, one in the Chatham Islands to the east of New Zealand, and the other on Snares and Solander islands to the south. Endemic to New Zealand.

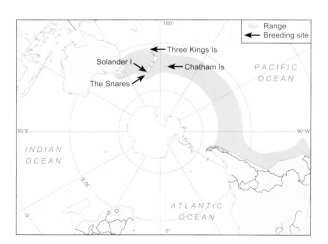

Description

An attractive and distinctive albatross, similar in size and general appearance to the other smaller mollymawks, but with a noticeably grey head, ashy forehead and black and yellow bill.
Identification: Distinguished from grey-headed albatross by lighter forehead and grey extending down sides of neck. Dark leading edge to underwing also more clearly defined. Juveniles have pale bill with dark

tip, versus all dark bill in grey-headed. Differs from Chatham and Salvin's by contrasting black and yellow bill and much broader dark leading edge to underwing, and paler bill in juveniles.

Adult male and female
Sexes similar.
Body and rump: White. Ventral region and undertail coverts also white. Mantle dark.
Head and neck: Uniform mid-grey but with pale grey forehead and crown. Eye dark with dark triangular eye patch and white crescent behind. Vivid orange gape line, visible only when displaying or feeding chick.
Bill: Near black with bright yellow-orange upper and lower ridges and tip. Bright orange skin at base of lower mandible.
Wings and tail: Upper wing all dark including mantle. Underwing white with broad dark leading edge and narrow dark trailing edge, primaries dark. Tail black.
Subspecies: *T. b. platei* (northern Buller's) generally darker grey head and neck and has slightly broader bill.

Juvenile
Similar to adult but grey head more diffuse and bill rather dull yellow with dark tip.

Size
Males are in general somewhat larger than females.
Body length: 76–81 cm.
Wingspan: 2.05–2.13 m.
Weight: Male 2.5–3.3 kg. Female 2.2–2.8 kg.
Bill: Male 117–129 mm. Female 113–124 mm.
Egg: Length 93–112 mm. Diameter 61–70 mm.

Population and distribution
Total numbers in the region of 32,000 annual breeding pairs, split into two distinct populations. Some 60 per cent (19,000 annual breeding pairs) are the northern Buller's subspecies *T. b. platei*, breeding on The Forty-Fours and The Sisters in the Chatham Islands, with 20 pairs also nesting on the Three Kings off the northern tip of New Zealand. The southern Buller's, *T. b. bulleri*, comprises about 13,600 annual breeding pairs in The Snares and Solander groups.

Oceanic range
Non-breeding adults and juveniles regularly frequent the South Pacific and Tasman Sea, ranging from the west coast of Australia across to South America, and up the coast to Peru. Breeding adults largely remain in the Tasman Sea and New Zealand territorial waters.

Breeding
Birds return mid-December on Snares and Solander with first eggs laid during first half of January. However, the northern populations on Chathams and Three Kings breed two to three months earlier. This was one argument in favour of separating them into distinct species. The nest is a typical mollymawk pedestal of mud, vegetation and debris, which is used on an annual or near-annual basis. Nesting habitats vary between the two subspecies: sites on Snares and Solander are located on tussock slopes, cliffs and under scrub and forest canopy, whereas in Chathams they are on open rocky ground and cliff tops.
Courtship: Very similar to all the other mollymawks, generally quite vocal.
Laying: Single white egg with red speckles mostly at larger end laid during early January on Snares and Solander and late October in Chathams and Three Kings.
Incubation: Averages 69 days, with chicks hatching in mid-March/early April in Snares and Solander, January and February in Chathams and Three Kings.
Fledging: Brooding period 15–30 days, a surprisingly large variation. Fledging after 167 days on average, with chicks departing in early September from Snares and Solander and June/July from Chathams and Three Kings.

NORTHERN BULLER'S ALBATROSS, CHATHAM ISLANDS

SOUTHERN BULLER'S ALBATROSS, THE SNARES

BULLER'S ALBATROSS, THE SNARES

Breeding success: 55–60 per cent, with 90 per cent of intact pairs breeding the following season.

Food
Surface feeder, largely small fish, squid, other cephalopods and crustaceans. Also a frequent scavenger around fishing vessels. Capable of diving to depths of a metre or more in pursuit of food.

Threats
Fishing: Commonly caught as bycatch due to scavenging habits; however, these losses have been reduced by the mitigation measures taken by New Zealand-registered fishing vessels.
Predation: Introduced weka, *Gallirallus australis*, an aggressive flightless bird, on Big Solander is a possible threat as it may take eggs and chicks. Predation of eggs and chicks by red-billed gulls, *Larus novaehollandiae*, subantarctic skua, *Catharacta lonnbergi*, and northern giant petrel, *Macronectes halli*. Danger of continuing human predation in Chatham Islands.
Habitat loss: Nesting sites on the Chatham Islands may be degraded by severe storms.

Atlantic Yellow-nosed Albatross *Thalassarche chlororhynchos*

Alternative or previous names: Western yellow-nosed albatross, *Diomedea chlororhynchos*; Carter's albatross, *T. chlororhyna*, *T. eximia*.
First described: Gmelin, 1789.
Taxonomic source: Brooke (2004), Robertson and Nunn (1998).
Taxonomic note: *Diomedea chlororhynchos* (Sibley and Monroe 1990, 1993) has been split into *chlororhynchos* and *carteri* and transferred to the genus *Thalassarche* after Brooke (2004) and Robertson and Nunn (1998).
Origin of name: From the Greek *chlorox*, 'pale green', and *rhynchos*, 'bill'. Common name descriptive and from breeding range.
Conservation status: Endangered.
Justification: Very restricted breeding range, being found only on Gough Island and the three islands of Tristan da Cunha. Long-term studies at two colonies indicate a population decline of nearly 60 per cent over a 72-year period. While there are land predator dangers, the chief threat is thought to be commercial fishing operations.

Description
A noticeably slender but otherwise fairly typical mollymawk that breeds in the South Atlantic. Along with its Indian Ocean cousin, the Indian yellow-nosed, it is the only albatross with a yellow stripe running only along the top of the bill.
Identification: Pale grey head distinguishes it from Indian yellow-nosed which has a white or nearly white head, but juveniles are virtually indistinguishable. Yellow on

culmen terminates in rounded 'U' shape whereas on Indian yellow-nosed it ends in a 'V' shape, diagnostic only at close range although not 100 per cent reliable. Orange-yellow stripe only on upper ridge of bill distinguishes it from Buller's and grey-headed, which have matching stripes along both mandibles. Relatively slim bodylines, especially the tail, compared to other mollymawks. Juveniles have ebony-black bills, darker than all others apart from Atlantic yellow-nosed.

Adult male and female
Sexes similar.
Body and rump: White with black or near-black mantle, undertail coverts white.
Head and neck: Pale grey with distinct dark triangular eye patch. Fine white crescent behind dark eye. Bright orange gape line on cheek visible only during displays or when feeding chick.
Bill: Black but with bright orange-yellow upper ridge, or culmen ending in 'U' at the base, and peach-orange tip. Orange skin at base of lower mandible.
Wings and tail: Upper wing black with very little greying on the mantle. Underwing white with broad black leading edge, narrow black trailing edge, and black wingtip. Tail black.

Juvenile
Similar to adult, but lacks grey head, while dark leading edge on underwing is less well defined; bill all black.

Size
Along with the Indian yellow-nosed it is the smallest of the mollymawks, males being generally slightly larger than females.

Body length: 75 cm.
Wingspan: 2.0 m.
Weight: 1.8–2.8 kg.
Bill: 107–122 mm.
Egg: Length 91–99 mm. Diameter 60–65 mm.

Population and distribution
Breeds in the South Atlantic on the Tristan da Cunha and Gough islands group. The main population is on Tristan, numbering 21,000–36,000 annual breeding pairs, with smaller populations on its satellite islands of Inaccessible and Nightingale. A further 5000–10,000 pairs breed on Gough Island, although this is possibly an underestimation. While existing figures provide a total of around 26,000–46,000 annual breeding pairs, anecdotal evidence suggests that populations have declined in recent years. Calculating the population is fraught with more difficulty than with most species, given the precipitous nature of the terrain and the often inclement weather, with most nesting areas loosely spread in dense vegetation ranging from fern scrub to tall tussock grass. The wide spread of the current estimates reflects this problem.

Oceanic range
The species appears to range largely in the South Atlantic between 15 and 45 degrees south and is frequently observed off the coast of Brazil, Argentina and south-western Africa. During the breeding season it forages near the breeding islands. After breeding, it disperses so that it is common along the coast of Uruguay and Argentina and north off West Africa to 15 degrees south; occasionally found in Indian Ocean and Australia, with one bird found ashore at the Chatham Islands, New Zealand.

ATLANTIC YELLOW-NOSED ALBATROSS, TRISTAN DA CUNHA ISLAND

ATLANTIC YELLOW-NOSED ALBATROSS, GOUGH ISLAND

Breeding

On Tristan, Nightingale and Inaccessible islands, birds return to nest in August, but up to one month later on Gough, with pairs generally breeding two years out of every three. The nest is a typical mollymawk truncated cone and bowl made of mud, peat and vegetation, generally clustered in small groups or loose colonies often among vegetation, either beneath the island tree, *Phylica arborea*, or amid stands of the miniature tree fern *Blechnum palmiforme*. Nesting generally takes place below 300 metres (985 ft) elevation on Gough, but above 500 metres (1640 ft) on Tristan. Immatures first return after five to 12 years at sea, and begin breeding at six to 13 years of age.

Courtship: A typical mollymawk courtship, though with nests spaced slightly further apart; courtship involves more movement than with species nesting in densely packed colonies. Vocalisations are restrained, consisting mainly of discreet grunts, and the notable absence of prolonged gaping wails. Courting groups often assemble on specific ridges or open areas, from which pairs will walk or fly to chosen nesting sites.

Laying: Single white egg laid in September and October.

Incubation: 63–70 days, hatching in November and December.

Fledging: Brooding lasts some 20 days and fledging takes 130 days.

Breeding success: Around 69 per cent on both Tristan and Gough. Pairs raise, on average, one chick every 2.4 years.

Food

Similar to other mollymawks, being largely squid and other cephalopods mixed with some fish and crustaceans. Surface feeders known to scavenge around fishing vessels.

Threats

Fishing: Frequently observed scavenging close to commercial fishing operations, especially longlining. Along the coasts of both South America and Africa, it is known to suffer from associating with these fisheries.

Predation: Human predation is now minimal, as it is legally protected throughout its nesting range. On Tristan there is a threat from the introduced black rat, *Rattus rattus*. While the introduced house mouse, *Mus musculus*, preys upon Tristan albatross and Atlantic petrel chicks in the winter months on Gough, there is no evidence that it attacks yellow-nosed chicks present only in summer. Chicks and eggs are taken by subantarctic skua.

ATLANTIC YELLOW-NOSED ALBATROSS, GOUGH ISLAND

Indian Yellow-nosed Albatross *Thalassarche carteri*

INDIAN YELLOW-NOSED ALBATROSS WITH GREY-HEADED ALBATROSS IN BACKGROUND, PRINCE EDWARD ISLAND, PHOTO P. RYAN

Alternative or previous names: Eastern yellow-nosed albatross, *Diomedea chlororhynchos*. Carter's albatross, *Thalassarche chlororhynchos bassi*.
First described: Rothschild, 1903.
Taxonomic source: Brooke (2004), Robertson and Nunn (1998).
Taxonomic note: *Diomedea chlororhynchos* (Sibley and Monroe 1990, 1993) has been split into *chlororhynchos* and *carteri* and both transferred to the genus *Thalassarche* following Brooke (2004) and Robertson and Nunn (1998). Interestingly, Alexander (1928) placed it in the genus *Thalassarche*.
Origin of name: After Tom Carter, collector of the type specimen at Point Cloates, Northern Australia in 1900.
Conservation status: Endangered.
Justification: Very limited breeding range with the some 55 per cent of the breeding population being on Amsterdam Island where the population has declined significantly due largely to two infectious bacteria, *Pasteurella multocida* (commonly known as avian cholera), and *Erysipelothrix rhusiopathiae*; presently up to 100 per cent of chicks die as a result of these two diseases, while the adult population is affected by industrial fishing operations.

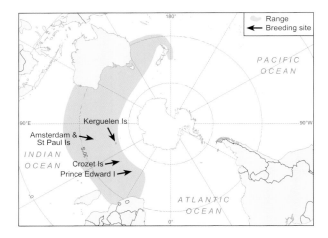

Description
A small slim black and white mollymawk, very similar to the Atlantic yellow-nosed, with which it was once taxonomically combined, but with a white or near white rather than grey head and found largely in the southern Indian Ocean.
Identification: Most easily confused with Atlantic yellow-nosed, which has a distinctly greyer head. Also, at very close range, the yellow on the culmen ridge has a rounded 'U' shape at the base, in contrast to the pointed 'V' terminal on Indian yellow-nosed, although this may not be 100 per cent diagnostic. Orange-yellow and black bill distinguishes it from shy, white-capped, black-browed and Campbell, but from grey-headed only by absence of yellow on lower bill. Black bill of juvenile differs from other mollymawks, but identical to Atlantic yellow-nosed.

Adult male and female
Sexes similar.
Body and rump: White with black or near black mantle, white ventral area and undertail coverts.
Head and neck: White with distinct dark triangular eye patch. Eye dark with white crescent behind. Bright orange cheek patch visible only when displaying or feeding chick.

INDIAN YELLOW-NOSED ALBATROSS, PRINCE EDWARD ISLAND, PHOTO P. RYAN

Bill: Black with bright orange-yellow upper ridge or culmen ending in a 'V' at the base. Bright orange skin at base of lower mandible.
Wings and tail: Upper wing black with pale primary shafts showing up in contrast. Mantle black, underwing white but leading edge has a broad dark band and trailing edge a narrow dark band. Wingtip black. Tail black and rather narrower in flight than other mollymawks.

Juvenile
Very similar to adult but some grey on head and nape forming a partial collar. Dark band on leading edge of underwing less well defined. Bill all black.

Size
Males are generally slightly larger than females.
Body length: 75 cm.
Wingspan: 2.0 m.
Weight: 2.5–2.9 kg.
Bill: 111–124 mm.
Egg: Length 93–100 mm. Diameter 58–62 mm.

Population and distribution
Breeds on four isolated island groups in the southern Indian Ocean, primarily on Amsterdam Island, with smaller colonies on Crozet and Prince Edward, and very small numbers on Saint Paul and Kerguelen. Some 55 per cent of the estimated total population of 36,000 pairs is located on Amsterdam. However, this is a 1998 estimate and in view of mortality caused by disease, the actual population could be in steep decline. Population figures for this island show a reduction from 37,000 pairs in 1984 to just 18,000 pairs in 2003. A single pair nests regularly among Chatham albatrosses on The Pyramid south of the Chatham Islands.

Oceanic range
Found throughout the southern Indian Ocean between 35 and 50 degrees south, extending east to Australia, the Tasman Sea and as far as New Zealand waters. Forages during breeding season close to breeding islands. During winter disperses to be abundant off southern Africa and western and south-eastern Australia, with birds occasionally reported from northern New Zealand.

Breeding
The nest is a typical truncated cone of mud, vegetation and other material which is reused on an annual basis. Nests are located in fairly close-knit colonies on tussock grass slopes and cliff faces. Adults return in late August or September.
Courtship: A typical mollymawk dance with bowing, fencing and mutual preening accompanied by occasional vocalisations.
Laying: A single white egg is laid in mid to late September and October.
Incubation: On average 71 to 78 days, hatching in late November and December.
Fledging: Averages 115 days with chicks typically fledging in late March or early April.
Breeding success: Averages around 25 per cent but near zero in some years due to bacterial infections.

Food
Similar to other mollymawks, surface feeding largely on fish and squid with some crustaceans. Also known to scavenge around fishing vessels.

Threats

Fishing: As a scavenger it suffers significant mortality as the result of commercial fishing activities.

Predation: Possibly taken by pinnipeds, sharks or giant petrels, otherwise subantarctic skua may prey on chicks and eggs. The presence of rats and cats on Amsterdam does not appear to have an impact.

Habitat loss: In the past, human-induced fire had a serious impact. In 1974 a major fire ranged through the colony, killing chicks on the nests.

Disease: Breeding success on Amsterdam Island, the main breeding location, is seriously affected by two bacteria, *Pasteurella multocida* (commonly known as avian cholera) and *Erysipelothrix rhusiopathiae*. With resulting high mortality in young chicks, and a lesser effect on adults, almost total breeding failure has been recorded in some years. While both diseases will respond to common antibiotics, administering such a treatment presents considerable difficulties, and as the pathogens can survive long periods in the open, it is hard to see how the situation can easily be improved.

INDIAN YELLOW-NOSED ALBATROSS, PRINCE EDWARD ISLAND. PHOTO P. RYAN

Laysan Albatross *Phoebastria immutabilis*

Alternative or previous names: Gooney bird. *Diomedea immutabilis*.
First described: Rothschild, 1893.
Taxonomic source: Brooke (2004), Robertson and Nunn (1998), Sibley and Monroe (1990, 1993), Stotz et al. (1996).
Origin of name: Common after main breeding island. *Immutabilis* meaning 'unchanging', reflecting the fact that juvenile and adult plumages are the same.
Conservation status: Vulnerable.
Justification: While current breeding population is quite large, the impact of commercial fishing has caused a significant decline in numbers which, left unchecked, would result in a further 30 per cent decline over the next 50–60 years.

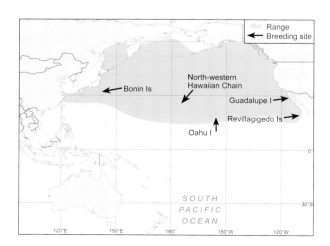

Description

A small albatross with black and white plumage, reminiscent of the mollymawks but found only in the North Pacific.

Identification: Dark mantle and upper wings, plus black patches on underwings distinguish it from the much larger short-tailed albatross, which also has yellow coloration on the head. Cannot be confused with any other albatross within its range.

Adult males and female

Sexes similar.
Body and rump: All white but with dark mantle which extends to the central part of the rump.
Head and neck: White, apart from dark patch around eye, and grey dusting on cheeks.
Bill: Peach, distinctly yellower at base, especially lower mandible; tip grey.
Wings and tail: Upper wing all black with pale primary shafts prominent. Underwing generally

white but with dark primaries and both leading and trailing edges black, plus irregular black patches in the central area. Mantle and tail black.

Juvenile
Same as adult, though upper rump patch may be larger. Bill and legs darker.

Size
Males somewhat larger than females.
Body length: 79–81 cm.
Wingspan: 1.95–2.03 m.
Weight: Male 3.00–3.60 kg. Female 1.95–2.4 kg.
Bill: Male 100–112 mm. Female 102 mm.
Egg: Length 100–121 mm. Diameter 62–72 mm.

Population and distribution
Breeds in 16 different locations in the subtropical North Pacific during the northern winter. Concentrated in the North-western Hawaiian Chain, the total annual breeding population is estimated to be around 437,000 pairs, with approximately 90 per cent nesting on Laysan Island and Midway Atoll (Sand, Eastern and Spit Islands). Smaller significant populations breed on Lisianski, Pearl and Hermes Reef, French Frigate Shoals and Kure Atoll, plus a few pairs on the islands of Kauai, Niihau and Oahu. Additionally no more than 500 pairs are split between Mukojima in the Bonin Islands off Japan, and on Guadalupe Island and the Revillagigedo Islands (Clarion and San Benedicto) off the coast of Mexico. The Mexican populations were first reported in the 1980s and may be a result of population expansion in the central North Pacific.

Oceanic range
With most breeding confined to the North-western Hawaiian Islands, birds either make very short foraging trips in nearby waters or move as far as the continental shelf off the Aleutian Islands. Non-breeders range throughout the northern Pacific, though not in inshore waters, tending to concentrate in the north-west Pacific in late summer.

Breeding
Breeding is annual, but 24 per cent of successful pairs take a year off, with all failed breeders returning the following year. Birds return in late October, nesting in open areas among grass and herbs, but not in thick scrub. Immatures return to nesting colonies at age of three, but breeding does not take place until aged eight or nine, tending slightly younger for males than females.
Courtship: Similar to the black-footed with which it shares a number of nesting locations. Actions include bill touching, tiptoeing, fake preening, and a 'sky groan' with various other vocalisations including a double 'eh-eh eh-eh'.
Laying: Single white egg laid in mid-November.
Incubation: Incubation lasts for 65 days, hatching in late January.
Fledging: Chicks are brooded for around 17 days. Fledging takes some 165 days, with birds leaving the colony in June or July.
Breeding success: Around 64 per cent, with 33 per cent of first-year birds and 82 per cent of second-year birds surviving.

Food
A surface feeder, reaching down to about one metre (3 ft), it preys mainly on cephalopods, fish, fish eggs and crustaceans. The eye of the Laysan possesses a high level of the night-vision pigment rhodopsin, comparable to barn owls and four times that of the diurnal-feeding black-footed albatross. This finding, along with a high proportion of squid in its diet, has given rise to speculations about it being a nocturnal feeder, yet studies using immersion monitors do not as yet corroborate this.

Threats
Fishing: Known to be at risk from commercial fishing. It was estimated that over 10,000 Laysan albatrosses were killed each year in longline and driftnet fisheries, but a worldwide industrial driftnet ban, and more recent longline bycatch

LAYSAN ALBATROSS, MIDWAY ATOLL

LAYSAN ALBATROSS, MIDWAY ATOLL

LAYSAN ALBATROSS, MIDWAY ATOLL

reduction techniques, have significantly reduced mortality levels.
Predation: Ruddy turnstones break unguarded eggs, while the bristle-thighed curlew (both winter migrants from the Arctic) has developed a technique for doing so using a stone held in its bill, though it is not thought that these have significant effects on the overall population. Up to 10 per cent of departing fledglings are taken by tiger sharks when they first alight near shore. Dogs, cats and rats kill chicks where nesting occurs on the inhabited islands of Hawaii.
Other: Additional threats come from oiling and ingestion of plastic at sea, and lead poisoning from decaying paintwork at certain breeding colonies close to buildings and decommissioned military establishments. Aircraft disturbance, once a significant problem, has been virtually eliminated since the instigation of night landings on breeding islands.

Black-footed Albatross *Phoebastria nigripes*

Alternative or previous names: Gooney bird, *Diomedea nigripes*.
First described: Audubon, 1849.
Taxonomic source: Brooke (2004), Cramp and Simmons (1977–94), Robertson and Nunn (1998), Sibley and Monroe (1990, 1993), Stotz et al. (1996), Turbott (1990).
Origin of name: Both common and scientific names refer to its black feet.
Conservation status: Endangered.
Justification: While still breeding in considerable numbers, population has shown steady decline and was predicted to drop a further 60 per cent over the next 50–60 years if measures are not taken to reduce mortality from longline fisheries.

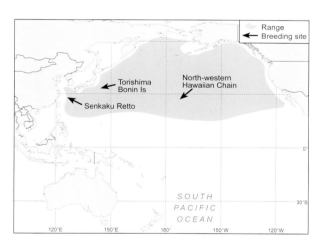

Description
A small dark-plumaged albatross found only in the North Pacific, north of the 20th parallel.
Identification: The only all-dark albatross in North Pacific, except for possible confusion with juvenile short-tailed albatross, though latter is larger and develops a pink bill and light-coloured feet and legs soon after fledging.

Adult male and femal
Sexes similar.
Body and rump: Dark charcoal grey, including mantle, mid-grey on rump, but whitish in older birds, along with undertail coverts.
Head and neck: Dark grey with diffused white 'ring' around the base of bill, shading gradually into dark facial plumage. Paler patch just behind eye.
Bill: All black.
Wings and tail: Uniformly dark black-brown, with shafts of primaries showing white.

Juvenile
Very similar to adult but with slightly browner tinge to plumage and no light patch near tail.

Size
Male generally somewhat larger than female.
Body length: 81 cm.
Wing span: 1.92–2.13 m.
Weight: Male 2.6–4.3 kg. Female 2.6–3.6 kg.
Bill: Male 102–113 mm. Female 94–110 mm.
Egg: Length 93–120 mm. Diameter 64–74 mm.

Population and distribution
The main breeding grounds are in the North-western Hawaiian chain. The total population is estimated to be around 54,500 annual breeding pairs. Of these some 95 per cent (±52,000) breed on Midway Atoll (Sand, Eastern and Spit Islands) and on Laysan Island. There are smaller populations on Kure Atoll, Pearl and Hermes Reef, Lisianski Island and French Frigate Shoals, with very small colonies on Necker, Nihoa and Kaula Islands. An additional three small colonies totalling some 2500 annual breeding pairs are located near Japan on Torishima, Bonin and Senkaku-retto.

Oceanic range
Throughout the North Pacific north of the 20th parallel, from Hokkaido in Japan to California, remaining at least 20–30 kilometres offshore. Much of the feeding is concentrated in the cold California Current along the west coast of North America, particularly during the non-breeding season, July through early November.

Breeding
In Hawaii, birds return to their nesting islands in early November, where nests usually consist of no more than a crater-like scrape in the sand, generally located in an open area, particularly near beaches, and spaced further apart than most other species. Where there is grass, as on Torishima, they will construct a modest structure. Immatures return to the colony at age three but will not breed until at least five years old. One bird was found still breeding at age 37, showing substantial longevity.
Courtship: In many respects more varied and faster paced than other albatrosses, consisting of fake preening, bill fencing, sky calling, and pirouetting by rising repeatedly on tiptoes, all accompanied by a variety of vocalisations including a double 'ha-ha, ha-ha'. The 'gooney walk', which has been described a sort of 'cakewalk', is not confined to courtship.
Laying: A single white egg laid in mid-November,
Incubation: 65 days, hatching in mid to late January.
Fledging: Chicks are brooded for around 19 days, with fledging after some 140 days, in June and July.
Breeding success: Estimated at just under 50 per cent, with roughly 6.9 per cent of eggs laid eventually producing breeding adults.

Food
Diurnal surface feeder, the black-footed albatross has limited nocturnal vision compared to Laysan albatross, which is thought to be a partially nocturnal feeder, and relies largely on flying fish eggs as well as squid and crustaceans. This ensures that the two species, whose ranges overlap, do not compete significantly for food.

Threats
Fishing: Commercial driftnet and longline fishing were considered to be the major cause of downward trend in the population. This has abated after the global banning of industrial driftnets in 1991, and with the implementation of longline bycatch mitigation techniques.
Predation: Previously harvested for its feathers, the species is now fully protected from direct human predation. Tiger sharks may take departing fledglings, and the bristle-thighed curlew uses a stone held in its bill to break some eggs.
Habitat loss: While habitat has been regained due to decommissioning of military bases at some prime breeding sites, many locations are very low lying and at risk from rising sea levels, wave action and tsunamis. On Torishima there is the risk of volcanic eruption.
Disease: Young are affected by avian pox virus on Midway Atoll, which causes some mortality.

Short-tailed Albatross *Phoebastria albatrus*

Alternative or previous names: Steller's albatross, *Diomedea albatrus*, *ahodori* (Japan).
First described: Pallas, 1769.
Taxonomic source: Brooke (2004), Robertson and Nunn (1998), Sibley and Monroe (1990, 1993), Stotz et al. (1996).
Origin of name: Specimen described by Pallas was collected by Georg Wilhelm Steller who travelled overland from Russia to Kamtchatka between 1738 and 1746. Common name a misnomer as tail is no shorter than others of the genus. *Ahodori* translates as 'fool bird'.
Conservation status: Vulnerable.
Justification: A very small breeding population of approximately 400 annual pairs found on two small islands south of Japan. While population has increased significantly over the past 20 years, it is still very small and additionally the main colony is on an active volcano. Incidental mortality in commercial fishing operations has been documented, but its impact on the population is unknown. Endemic to Japan.

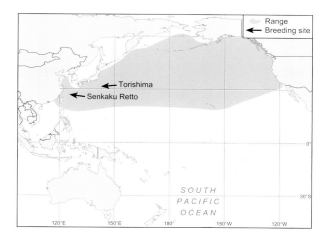

Description
A medium-sized albatross with a large pink bill and distinctive black and white plumage and rusty yellow nape. The feet normally extend beyond the end of the tail, which is not noticeably short. Found exclusively in the North Pacific.
Identification: The only albatross with a white body and mantle found in the North Pacific. Its pink bill and white on upper wing and mantle distinguish it from the adult Laysan albatross. The juvenile possibly confused with black-footed albatross but appreciably larger, and pink bill, developed very soon after fledging, contrasts with the dark bill of black-footed.

Adult male and female
Sexes similar.
Body and rump: All white.
Head and neck: Pale orange-yellow crown, neck and collar; face and chin white.
Bill: Large and pink with a bluish tip.
Wings and tail: Upper wing white with black primaries and primary coverts. Black irregular trailing edge to inner wing. Underwing white with black tip and narrow black trim along both edges. Tail white with broad black terminal band.

Juvenile
Almost entirely dark chocolate brown with distinctive large pink bill (dark coloured on fledging, but lost within first couple of months).

Immature
Intermediate plumage, largely brown with white vermiculated mantle, belly and breast, and dark crown and nape. Distinctive white wing patches. Becoming gradually whiter with age. Back of neck remains brown longest, giving the bird a slightly punk look.

Size
Males generally slightly larger than females.
Body length: 84–91 cm.
Wingspan: 2.13–2.35 m.
Weight: 5.1–7.5 kg.
Bill: 129–141 mm.
Egg: Length 111–125 mm. Diameter 70–79 mm. Weight 310–375 g.

Population and distribution
Breeds only on two islands, the main colony on Torishima south of Japan, and two small but growing colonies in the Senkaku-retto group in the East China Sea, north of Taiwan. The species was believed extinct by 1949, after the last breeding birds were killed in 1931, but a handful of mature birds returned in 1951 and restarted the colony with the first egg being laid in 1954. At the last available information, 341 nests were counted on Torishima in 2006/07, plus 85 were estimated on the Senkakus in 2005/06 (H. Hasegawa pers. obs.). With an estimated 80 per cent of breeders nesting each year, the current world population is

approximately 500 breeding pairs in total. This is a small fraction of numbers in the nineteenth century, when over five million adults were killed, mainly for their feathers. Efforts are under way to entice a new breeding colony in the Ogasawara, or Bonin, Islands through the use of decoys and sound recording. Two individuals have occasionally visited Midway Atoll in Hawaii, but the same methods have failed to bring them together as they settled in separate locations.

Oceanic range

Found mainly along the continental shelf margins throughout the Pacific Rim, above 25 degrees north. Remains primarily within Japanese waters during chick rearing. Post-breeders disperse widely, concentrating in the Alaskan and Russian waters, particularly around the Kuril and Aleutian Islands, and the Bering Sea. Non-breeders and juveniles in particular frequent the west coast of Canada and the US. Little time is spent in the central oceanic gyres, with aggregations along continental shelf breaks, including the heads of submarine canyons. On one occasion approximately 10 per cent of the world population was observed concentrated near St Matthews canyon in the Bering Sea.

Breeding

Birds return to Torishima in October. Immatures return at about three to four years, but do not start breeding until eight or nine, with the earliest recorded being five years. The nest is a low mound of vegetation, sand and volcanic debris up to 60 centimetres (2 ft) across. Records kept by Dr Hasegawa since the Torishima colony re-established in 1951 suggest a lifespan up to 45 years.

Courtship: Similar to the black-footed albatross including a sky call and 'bill-under-wing' or fake-preen posture. The most common vocalisations are loud wails and a double croak-like call 'eh eh'.
Laying: Single white egg laid late October and early November.
Incubation: 64–65 days, hatching in late December and January.
Fledging: 150–160 days with birds leaving the colony by mid-June.
Breeding success: In recent years 67 per cent.

Food

Surface feeder and scavenger, its main food is squid and other cephalopods, fish, including flying fish eggs, and crustaceans. Previously reported catching spawning salmon at the mouths of rivers on the Kamchatka Peninsula.

Threats

Now fully protected, it remains highly vulnerable due to the small population.
Fishing: As a scavenger, it is vulnerable to bycatch in North Pacific fisheries. Protection in most Pacific Rim countries with exclusive economic zones frequented by this species has raised awareness within respective governments. Stringent mitigation regulations can reduce fisheries mortality to near zero; however, such requirements are not in place throughout the species range.
Predation: No known predators; black rats are common on Torishima, but apparently do not take eggs or chicks.
Habitat loss: As much of the nesting area is relatively unstable volcanic material, tussock and other grasses have been planted to help stabilise the substrate and prevent egg loss through erosion. Torishima is an active volcano with periodic violent explosions. In 1902 and again in 1939 the nesting area was buried by eruptions, though luckily not during the nesting season. This threat remains and has prompted efforts to establish new colonies at less vulnerable sites, successfully achieved through the use of decoys and sound playback.

SHORT-TAILED ALBATROSS, MIDWAY ATOLL

SHORT-TAILED ALBATROSS, MIDWAY ATOLL

Waved Albatross *Phoebastria irrorata*

Alternative or previous names: Galapagos albatross. *Diomedea irrorata*.
First described: Salvin, 1883.
Taxonomic source: Brooke (2004), Robertson and Nunn (1998).
Origin of name: *Irrorata* means 'sprinkled' or 'specked'; both Latin and common names referring to the fine wavy markings on the breast and flanks.
Conservation status: Critically Endangered.
Justification: Status has recently been uplisted due to new information and circumstance. Breeds on just a single small 14-square kilometre (5.4-sq mile) island of Galapagos in the tropical eastern Pacific. While fully protected, its nesting habitat has changed as an aftermath of feral goat eradication. Recent reduction in life expectancy appears to reflect increased pressure from commercial fishing on main feeding grounds in Peru, with known targeting as a food source by artisanal fishermen. Endemic to Ecuador.

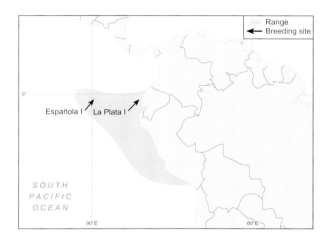

Size
Males notably larger than females.
Body length: 90 cm.
Wingspan: 2.30–2.40 m.
Weight: Male 3.40–4.00 kg. Female 2.90–3.20 kg.
Bill: Male 142–160 mm. Female 134–157 mm.
Egg: Length 106 mm. Diameter 70 mm (av.). Weight 284 g. (av.), larger for older females.

Population and distribution
The only viable breeding population is on Española Island in the Galapagos Archipelago off Ecuador, estimated at 15,000–18,000 annual breeding pairs. Recently, small numbers of birds have been observed on Genovesa Island in the north of the archipelago, but no breeding has resulted. An insignificant number frequent Isla de la Plata

Description
The only truly tropical albatross, it has the smallest range in the family. Unmistakable, it is of medium size, with an unusually long, bright yellow bill.
Identification: Distinguished from several mollymawks whose ranges overlap off the coast of Peru by its brown rather than black wings, plus darker body and rump. The yellow head colouration is also diagnostic, as is its much longer bill.

Adult male and female
Sexes similar.
Body and rump: Above mid-brown with paler, slightly barred rump. Below light grey-brown becoming grey and white with fine wavy vermiculation on breast and flank.
Head and neck: White with delicate buff/gold on head and back of neck. Distinctive, jutting eyebrows feathers give it a 'square-eyed' look.
Bill: Uniform dull golden yellow, but with darker, slightly greenish-tinged tip, notably hooked, heavier in male.
Wings and tail: Upper wing dark brown, underwing cream-buff coloured with dark leading and trailing edges. Tail brown.

Juvenile
Lacks subtleties of plumage of adult; bill pale yellow.

near the Ecuadorian coast, although no successful breeding has been recorded. Recent research suggests that average adult life expectancy has dropped abruptly from 20.3 years down to just 12, presumably due to fisheries impact, a reduction that could have catastrophic consequences for a species whose age at first breeding averages five years.

Oceanic range

The species has the most restricted range of any albatross. It is seen regularly throughout the Galapagos Marine Reserve in the breeding season, where it feeds during the chick guarding period. Otherwise, forages mainly to the south and east of the islands in the upwelling zones along the coast of Peru, where it also spends the non-breeding season.

Breeding

The breeding period coincides with the 'garua' or cool season when the south-east trade winds blow, with adults returning in late March and all birds departing by late December. Males are very aggressive and may copulate with any female available. Latest studies indicate that some 25 per cent of chicks are fathered by another male. Normally, no nest is built, the egg being incubated on the bare ground and shuffled considerable distances held between clamped tarsi, a behaviour unheard of in any other bird except emperor and king penguins. Loose colonies located in scrub and more open ground close to cliff edge on southern, windier side of island. First breeding is at four to six years, although several more years may pass before they successfully raise their first chick.

Courtship: An elaborate ritual, it starts with 'billing' or bill touching, and involves fencing, bowing, sky-pointing accompanied by a soft moo, gaping, bill circling and bill clacking/rattling, as well as a formalised dance or 'sway walk' with exaggerated side-to-side head movement. Intense displays can be prolonged.

Laying: A single white egg is laid on bare ground between mid-April and late May.

Incubation: The mean incubation period is 65 days. During incubation the egg may be moved 40 metres (130 ft) or more. Hatching in mid-June through mid-July.

Fledging: Average fledging period is 179–188 days. Chicks wander in search of shade, parents calling for them on return from foraging trips. Immatures fledge late November and December.

Breeding success: Highly variable, in the range 4–82 per cent, due largely to climatic factors, such as heavy rains and mosquitoes or high temperatures/humidity during El Niño years. Immature survival rates around 80 per cent compared to 90 per cent in adults.

Food

Mainly fish and squid from surface feeding, with occasional shallow lunge dives. Not normally a ship follower, yet attracted to offal from fishing vessels in Peruvian waters where it may take individually baited hooks. Known to harass boobies for food and to feed in conjunction with dolphins.

Threats

Fishing: While the waved albatross has a very limited range, it is at risk from commercial fisheries, with evidence from band recoveries further indicating it is targeted as food by artisanal fishers off the coast of Peru.

Predation: Galapagos hawks take small numbers of chicks; Española mockingbirds will attack unprotected eggs, often in small groups, although they are rarely able to pierce it unless already damaged.

Habitat loss: There is some indication that after the removal of feral goats from Española in the 1980s vegetation regrowth has reduced the area suitable for nesting.

WAVED ALBATROSS, ESPAÑOLA ISLAND, GALAPAGOS

WAVED ALBATROSS, ESPAÑOLA ISLAND, GALAPAGOS

WAVED ALBATROSS, ESPAÑOLA ISLAND, GALAPAGOS

Light-mantled Albatross *Phoebetria palpebrata*

Alternative or previous names: Black-billed albatross, light-mantled sooty albatross, quakerbird. *Phoebetria fuliginosa*.
First described: Forster, 1785.
Taxonomic source: Brooke (2004), Christidis and Boles (1994), Dowsett and Forbes-Watson (1993), Robertson and Nunn (1998), Sibley and Monroe (1990, 1993), Stotz et al. (1996).
Taxonomic note: Originally considered as a single species together with the sooty albatross, the division was not clarified until the end of the nineteenth century.
Origin of name: *Palpebra* Latin for 'eyelid', referring to white crescent above and behind the eye.
Conservation status: Near Threatened.
Justification: Although widespread and breeding on several isolated islands in the southern oceans, it is believed to suffer mortality from longline fisheries. The status of the breeding populations is uncertain, but its very slow reproduction rate and indications of population decline are cause for concern.

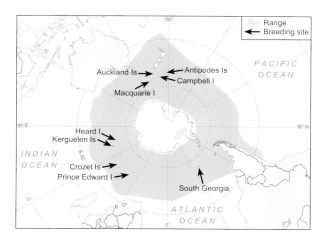

Description
An elegant, all-dark albatross with slender, pointed wings and a long diamond-shaped tail, similar physiologically to the sooty albatross, but ranging much further south.
Identification: With the sooty, the only altogether dark albatross in the southern hemisphere, distinguishable by its pale mantle. Slender wings and tail set it apart from dark morph giant petrels.

Adult male and femals
Sexes similar.
Body and rump: Sooty brown to dusty mid-grey with distinctly lighter mantle.
Head and neck: Dark chocolate-brown face paling into light grey nape and faded mantle. Eye dark with white crescent above and behind it.
Bill: Black but with pale blue stripe on sulcus along sides of lower mandible.
Wings and tail: Upper wing charcoal-grey or brownish with primary shafts showing white. Underwing appears somewhat darker brown. Tail dark brownish-black, long and diamond shaped, undertail coverts dark grey.

Juvenile
Virtually identical to adult, though may have dark scalloping on back and lacks both blue on sulcus and white eye crescent.

Size
Males normally somewhat larger than females.
Body length: 78–90 cm.
Wingspan: 1.80–2.20 m.
Weight: Male 2.8–3.1 kg. Female 2.6–3.7 kg.
Bill: Male 103–117 mm. Female 98–117 mm.
Egg: Length 97–103 mm. Diameter 63–67.5 mm.

Population and distribution
Major populations throughout the subantarctic islands, on South Georgia, Crozet, Kerguelen, Macquarie, Auckland and Campbell Islands, with smaller numbers on Marion, Prince Edward, Heard and Antipodes islands. The total annual breeding population is estimated at 19,000–24,000 pairs.

Oceanic range
Widespread throughout the southern oceans, but generally staying near or below the Antarctic Convergence. Appears to forage in rather higher

LIGHT-MANTLED ALBATROSS CHICK, CAMPBELL ISLAND

latitudes than other albatrosses, regularly venturing down to the edge of the pack ice.

Breeding

A biennial breeder, it nests individually or in small groups on a number of isolated oceanic islands in the southern oceans, selecting cliff ledges, generally coastal rather than inland, but often at some considerable height above the sea. Although sharing some of the same Indian Ocean islands as the sooty albatross, it tends to be more dispersed and occupies steeper terrain. The nest is a low pedestal similar to most southern albatrosses, made of mud and vegetation up to 20 centimetres (8 in) high. Birds return to breeding islands in September or early October depending on locations. First breeding is age seven to eight, though this can vary, with some females starting as early as four years. Records from Crozet show adults still breeding aged 32.

Courtship: Quite similar to the sooty, with considerable time spent in aerial courtship, and with the same terrestrial displays including the haunting 'Pee-oow' vocalisation uttered while sky-pointing.

Laying: A single white egg with irregular red blotches at the larger end, laid late October in a concentrated 10–12 day period.

Incubation: Normally around 70 days, hatching in early January.

Fledging: Brooding period lasts for around 20 days, with fledging taking between 140 and 170 days depending on locality, departing in June and July.

Breeding success: In the region of 20 per cent but varying with location and year. On Macquarie pairs raise one chick every three or four years, thought to be the lowest reproductive rate of any bird species.

Food

Squid, other cephalopods and fish are the most important food source, with crustaceans and carrion also taken, the last including penguins and small petrels. Little is known of feeding habits, but there is evidence to indicate that they can dive to as much as 12 metres (39 ft), and with their oddly shaped, long-pointed fleshy tongues it seems likely that there is a lot to learn about the feeding behaviour of both the light-mantled and sooty albatrosses.

Threats

Fishing: While commercial fishing activities are believed to impact this species, particularly long-lining operations, there is no clear information as to the severity of the problem.

Predation: Feral cats were predators on Macquarie until eradication in 2001, and possibly on Campbell, where they have now vanished. Giant petrels are known to attack adults as well as chicks. Skuas and sheathbills may also prey on eggs and chicks.

LIGHT-MANTLED ALBATROSS, ANTIPODES ISLAND

Sooty Albatross *Phoebetria fusca*

Alternative or previous names: Peeoo (local on Tristan da Cunha); Quakerbird. *Diomedea fusca.*
First described: Hilsenberg, 1822.
Taxonomic source: Brooke (2004), Christidis and Boles (1994), Dowsett and Forbes-Watson (1993), Robertson and Nunn (1998), Sibley and Monroe (1990, 1993), Stoltz et al. (1996).
Taxonomic note: While originally included in the same genus, *Diomedea*, as other albatrosses, they were placed in a separate genus, *Phoebetria*, by Reichenbach in 1853 where they have remained in spite of the revolution in albatross nomenclature following Robertson and Nunn (1998).
Origin of name: From Greek *Phoebus* (Apollo), 'The Bright One', God of light, or *Phoebe* (Diana) and *betria*, inspired or prophetic. *Fusca* refers to the dusky plumage, as does the common name.
Conservation status: Endangered.
Justification: Modest population with 66 per cent nesting on Tristan da Cunha and Gough Island group, where populations show evidence of serious decline (58 per cent in 15 years) and on Marion (25 per cent in eight years). This loss is likely due to a combination of commercial fishing, predation and avian diseases. A biennial breeder, the species has a very low reproductive rate. Relatively little information on the species is available, in part due to its cliff-nesting habit which makes access and conducting censuses difficult.

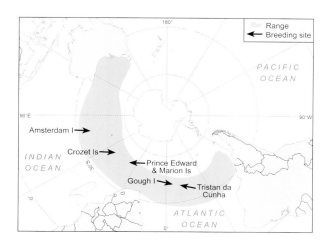

Description
A distinctive, elegant and entirely dark albatross, noticeably more slender and streamlined than other albatrosses, apart from the light-mantled of the same genus. The tail is long and diamond-shaped, which accounts for its aerial agility and its amazing aerial courtship displays.
Identification: Distinguished from light-mantled albatross by darker overall plumage and yellow bill stripe instead of pale blue. Distinct from giant petrel by slimmer shape and more slender, dark bill.

Adult male and female
Sexes similar.
Body and rump: Uniform dark charcoal grey to chocolate-brown.
Head and neck: Same as body, with slight fading on nape and sides of head, eye dark with dramatic white crescent above and behind it.
Bill: Black and more slender than other albatrosses, but thicker and straighter than the light-mantled. Distinctive mustard-yellow stripe on the sulcus along the sides of lower mandible.
Wings and tail: Long and narrow, almost black, slightly darker than body and head. White primary shafts noticeable in flight. Tail nearly black, long and wedge- or diamond-shaped, especially when fanned in courtship.

Juvenile
Similar to adult, bill stripe and eye crescent grey, with brown shafts to primaries.

Size
Males slightly larger than females.
Body length: 84–89 cm.
Wingspan: 2.05 m.
Weight: Male 2.20–3.25 kg. Female 2.10–2.80 kg.
Bill: Male 109–117 cm. Female 101–112 cm.
Egg: Length 95–108 mm. Diameter 59–69 mm.

Population and distribution
Breeds on isolated islands primarily in temperate or subtropical regions of the South Atlantic and Indian oceans, with the main population concentrated in the Tristan da Cunha group and Gough Island. Smaller populations are scattered in the Indian Ocean on Prince Edward and Marion islands, the Crozet group and on Amsterdam, plus a few pairs

on Kerguelen. Annual breeding pairs number approximately 12,500 to 19,000.

Oceanic range
Generally in temperate and subtropical waters in the South Atlantic and Indian oceans as far east as Tasmania, north to about 20 degrees south; rarely extending to 64 degrees south in the Indian Ocean, and as vagrants around New Zealand subantarctic islands and Macquarie Island. Feeding range tracked during the breeding season is around 600 kilometres (373 miles) from nesting island.

Breeding
In contrast to most other albatrosses, breeds in small groups on cliffs, both coastal and inland, which makes study difficult. Birds return in late September. The nest is a typical albatross pedestal with a deep bowl, but not as large or durable as many mollymawks. Immatures remain at sea for at least five years, with first breeding at nine years, or as late as 15 years.
Courtship: Dramatic and evocative aerial courtship displays, with pairs flying in unison and mirroring each other's every movement. Their terrestrial display is less varied than other albatross species, but includes sky calling, bill fencing and snapping, tail fanning and twisting, and mutual preening. The sky call is accompanied by a haunting 'Pee-oow' addressed to birds in flight, especially evocative when heard from a misty ledge high in an inland valley or glen.
Laying: A single white egg with reddish brown spots at broad end, laid in early October.
Incubation: 70 days, hatching in mid to late December.
Fledging: Chicks brooded for around 21 days, fledges at 145 to 178 days depending upon location, leaving in late April through mid-June.
Breeding success: Varies according to location and year, anything from 10 per cent to 50 per cent, with failures due to a variety of causes, particularly desertion, although the reasons remain unknown. Older birds appear to have a better success rate than new breeders.

Food
Feeds mainly on squid and other cephalopods, fish and crustaceans, but also on carrion, including penguins and petrels. Generally thought to be a surface feeder, but few observations exist, possibly due to nocturnal or twilight feeding. There is no evidence that it dives as deep as the light-mantled albatross, but this may be due to lack of information.

Threats
Fishing: Not a major scavenger behind fishing vessels, but is believed to be at risk from longline fisheries.
Predation: Some risk from feral cats on Amsterdam. Also, being smaller than other species, there is some evidence of giant petrel predation. Skuas and sheathbills may also prey on eggs and chicks and account for relatively low breeding success rate.
Habitat loss: Some degrading of nesting sites by free-ranging cattle on Tristan.
Disease: May be affected by avian cholera and the *Erysipelas* bacteria on Amsterdam.

Where to See Albatrosses

There are not very many places around the world where it is possible to see albatrosses on land, so remote are most of their nesting grounds. To see them courting and dancing, or tenderly taking care of their chicks, is a moving experience indeed. Still, it should be remembered that for anyone to begin to claim to know an albatross one must have travelled with them far, far away from land. Gliding over the ocean, whether off the coast of Peru, Alaska, Japan or New Zealand, or riding the ferocious storms of the great Southern Ocean, this is where their spirit truly soars, as free as the wind.

Galapagos
Tour itinerary must include Española Island.
Season: Late April to November, best May–June.
The Galapagos waved albatross has the most restricted range of any species, yet is also one of the easiest in the world to see, a startling sight amid cacti and marine iguanas. Many options available.

Hawaii, Midway Atoll
Visitor access is being developed at this writing.
Season: November to June, best February–March.
A total albatross immersion among one million Laysan and 50,000 black-footed albatrosses, plus many other seabirds.
www.fws.gov/midway/intro/default.htm

California
Pelagic tours in Monterey Bay.
Season: Autumn/winter/spring.
Major feeding area for black-footed albatross, with some Laysan and, exceptionally, a juvenile short-tailed albatross. www.shearwaterjourneys.com
www.montereyseabirds.com

Alaska
At-sea birding.
Season: Spring/summer.
Black-footed albatross are common off the south coast, while Laysan frequent waters around the Aleutian Islands, with rare sightings of short-tailed albatross near the end of the island chain around Attu. www.Ventbird.com

Falkland Islands
Cruising en route to Antarctica or self-catered on a private island.
Season: December to March.
There are several large black-browed albatross colonies, some of which are visited by vessels cruising to Antarctica. On several islands, reached by local flight, it is possible to book private lodging within walking distance of the nesting colony.
http://wikitravel.org/en/Falkland_Islands

South Georgia
Classic subantarctic island.
Season: November to March.
Best place to see the wandering albatross nesting on Prion Island, reachable only by cruise ship or private yacht charter. Light-mantled and grey-headed are also present but rarely seen up close.

New Zealand
Albatross capital of the world, with half of all species nesting.
Taiaroa Head, Dunedin: The small colony of northern royal albatrosses can be seen from the observatory and guided tour. www.albatross.org.nz
Kaikoura (all year, best June–August): One of the best places to see albatrosses at sea up close and personal. Eight species regularly, including most New Zealand endemics. Albatross Encounters runs small, intimate tours daily.
www.oceanwings.co.nz/albatross/
Subantarctic islands (November–February): Amazing albatross diversity on five island groups. Landings are permitted only on Campbell (southern royals and light-mantled) and Auckland (white-capped), with near-shore cruising at the others (Buller's and Salvin's). www.heritage-expeditions.com

South Africa
Day trips from Cape Town.
Season: all year, more variety April–September.
Great convergence area for rarely seen types like the sooty, and both Indian and Atlantic yellow-nosed, plus wanderers, grey-headed and shy from Australia–New Zealand. All proceeds from Cape Town Pelagics are donated to the Birdlife Albatross Campaign. www.capetownpelagics.com/

Indian Ocean
French logistics vessel servicing subantarctic bases.
Season: November, December and March trip dates.
Passage can be booked on the resupply vessel to Crozet, Kerguelen and Amsterdam islands. Although albatross visits aren't assured, this is a very unusual destination. www.taaf.fr (French islands)

Australia
Pelagic day trips with Southern Ocean Seabirds Study Association.
Season: June to November
Sailing from Wollongong, south of Sydney, with eight species seen regularly, plus many other rare sightings (18 species recorded), including one sighting of critically endangered Amsterdam albatross. www.sossa-international.org

Glossary: Selected Terms and Abbreviations

Mark Jones

Throughout this book it is our intention to educate and broaden perspectives on the realm of albatrosses and, indeed, make clear the connections between their world and ours. We have endeavoured wherever possible to use plain language and make the text readily accessible to non-specialist audiences. However, in a body of work spanning such a broad variety of professional and scientific disciplines, it is inevitable that we introduce the reader to terms and concepts that are beyond the normal lexicon of everyday language. We hope the following explanations and additional background information will prove helpful in clarifying and furthering the understanding of the specialist topics discussed.

ACAP Agreement on the Conservation of Albatrosses and Petrels Developed since 1999 under the auspices of the Convention on the Conservation of Migratory Species of Wild Animals (CMS), it is a multilateral agreement which seeks to conserve albatrosses and petrels by coordinating international activity to mitigate known threats to albatross and petrel populations. The Agreement entered into force on 1 February 2004, and to date there are 11 Party nations: Argentina, Australia, Chile, Ecuador, France, New Zealand, Norway, Peru, South Africa, Spain and the United Kingdom (including its Overseas Territory of Tristan da Cunha). Brazil is a signatory but has yet to ratify.

Antarctic Convergence, also known as the Antarctic Polar Frontal Zone (abbreviated to 'Polar Front') A zone encircling Antarctica (varying somewhat in latitude) where the cold, northward-flowing Antarctic waters sink beneath the warmer waters of the *subantarctic*, creating a nutrient-rich region where abundance of marine life congregates. The zone is approximately 32–48 kilometres (20–30 miles) wide, extending across the Atlantic, Pacific and Indian Oceans between the 48th and 61st parallels of south latitude.

anthropogenic Caused by human activity.

archipelago An extensive group of scattered islands.

Argos (also ARGOS) Advanced Research and Global Observation Satellite A unique worldwide location and data-collection system dedicated to studying and protecting the environment. Created in 1978 by the French Space Agency (CNES), the US National Aeronautics and Space Administration (NASA) and the US National Oceanic and Atmospheric Administration (NOAA), and now with active participation from several other international space agencies, thousands of animals fitted with miniaturised transmitters (see also PTT) can be tracked anywhere they travel on the planet.

artisanal fishery Near-shore subsistence and/or small-scale commercial fishing, usually by owner-operator 'household' crews making short duration trips, for local markets. Can refer to vessels from one-person canoes employing purely traditional fishing techniques to modern, well-equipped *trawlers* of 20+ metres (70 ft), seiners or *longliners*. In reality, heavy investments and developing techniques, together with broader marketing of product, can blur the definition of the contemporary artisanal fisher.

ATS Antarctic Treaty System, whose primary objective is to ensure 'in the interests of all mankind that Antarctica shall continue forever to be used exclusively for peaceful purposes and shall not become the scene or object of international discord'. In effect since 1961, it currently has 46 member nations.

avian Of or relating to birds, members of the *taxonomic* class Aves.

band, banding Sometimes also called ringing; universal method of permanently tagging individual birds with specially designed numbered and/or coloured leg bands, for long-term study and future identification. Permits and rules are governed by national and international bodies, which also maintain the detailed databases. The oldest known banded bird was 'Grandma', a northern royal albatross from Taiaroa Head in New Zealand: banded in 1937, her life history was followed for over 60 years.

BAS British Antarctic Survey The UK's national Antarctic operator, and has for the past 60 years been responsible for most of the UK's scientific research in Antarctica. Part of the Natural Environment Research Council (NERC), BAS operates five research stations, two Royal Research Ships and five aircraft in and around Antarctica.

biennial Occurring every two years — as opposed to biannual which means twice per year. Many albatrosses only breed biennially at best.

BLI BirdLife International A global partnership of conservation organisations, operating in over 100 countries and territories worldwide, that strives to conserve birds, their habitats and global biodiversity, working with people towards sustainability in the use of natural resources.

brooding Refers to a bird sitting on eggs to incubate them; also used during the 'guard-stage' when parents sit on young chicks to protect them from predation and the elements.

bycatch or by-catch During fishing operations, the part of a catch, representing all species living or dead, taken incidentally and in addition to the target species towards which the fishing effort is directed.

CCAMLR Commission for the Conservation on Antarctic Marine Living Resources Came into force in 1982, as part of the Antarctic Treaty System. With 24 Member States and 10 Acceding States, the aim of the Convention is to conserve marine life of the *Southern Ocean*.

cephalopods The class of bilaterally symmetrical 'head-footed' marine molluscs featuring large brains and well-developed senses that include all octopus, squid, cuttlefish and nautilus. An ancient group of animals comprising about 800 species, they are found throughout the world's oceans.

CITES Convention on International Trade in Endangered Species of Wild Fauna and Flora An international agreement between governments whose aim is to ensure that international trade in specimens of wild animals and plants does not threaten their survival. Currently it has 172 Contracting Parties.

cohort In biology and demographics, a group of animals born during the same period or breeding season; a generational group.

continental shelf The relatively shallow underwater outer margins of a continent, usually a plateau comprising the seabed and substrate, extending to where there is a marked increase in slope towards the oceanic depths. Conventionally considered as ranging from 0 to 200 metres (0–656 ft) in depth, these waters generally support the richest marine life concentrations. Most of this area falls within the jurisdiction of individual nations' *Exclusive Economic Zones* (EEZ).

crustaceans Group of chiefly aquatic arthropods with external skeletons, comprising some 40,000 species including crabs, lobsters, shrimp, crayfish and barnacles (but also terrestrial hermit crabs and woodlice!).

culmen or culminicorn A distinctive feature of a bird's bill or beak, it is the upper central ridge of the top mandible, often referred to as a key to species identification.

demersal fishery Fishing techniques that target fish stocks living near the sea floor (see also *longlining* and *trawl fisheries*).

demographic Related to the study of population statistics and trends.

Diomedeidae The *taxonomic* family that comprises all albatross species; one of four families within the defining order *Procellariiformes* representing all the 'tube-nosed' seabirds.

diurnal Opposite of nocturnal; refers to an animal that is active by day.

DoC New Zealand Government Department of Conservation, the body responsible for the conservation and management of main breeding grounds of nearly half all albatross species.

driftnet fishery Fishing techniques employing large (up to 50 kilometres [30 miles] long) gillnets, held vertically in the water between floats and weights, that drift freely with the currents. In 1991, as a result of controversial bycatch levels, the United Nations National Assembly called for a worldwide ban on the *high-seas* use of nets longer than 2.5 kilometres (1.5 miles).

dynamic soaring Flying technique requiring minimal effort by exploiting wind gradients and wave patterns to maintain forward momentum, producing a characteristic see-sawing flight pattern.

ecosystem A community of all living organisms, the delicate interactions between them and their relationships to the physical environment they inhabit. The concept is applicable on any scale and, since no system is closed, precise practical definition is difficult: can equally relate to a pond, an ocean or the entire planet.

EEZ Exclusive Economic Zone Under the United Nations Convention of the Law of the Sea this is the area in which a coastal state has sovereign rights to exploit – as well as the responsibility to conserve and manage – all the economic resources of the sea, seabed and subsoil, usually from the limit of Territorial Waters (12 nautical miles) out to 200 nautical miles.

El Niño A periodic disturbance in regular weather patterns and ocean currents along the Ecuador-Peru coast; the term means 'the child' in Spanish since it manifests itself around Christmas. Features abnormally warm waters and exceedingly high rainfall, with an associated catastrophic crash in productivity of the affected marine ecosystem, may cause enormous mortality among the region's seabird populations. Widespread climatic anomalies in other regions can be attributed to the global effects of El Niño.

endemic, endemism An organism that is native to and confined within a specific region; isolated islands especially have evolved species that are highly restricted in their distribution.

FAO Food and Agriculture Organization of the United Nations Since 1945, a neutral international forum for negotiating agreements, debating policy and sharing expertise, and collecting and disseminating information, with the aim of helping developing countries modernise to improve food production practices such as agriculture and fisheries.

fecundity In ecology and *demographic* studies, it is the potential reproductive capacity of an organism or population, which has been shown to increase or decrease in response to current conditions and regulating factors, e.g. climate, food supply, etc.

fledge, fledging Ready to fly, as in a young bird that has fully developed feathers and is ready to become independent, i.e. becoming a fledgling.

gam, gamming Term derived from whalers or seafarers visiting or meeting each other at sea ('It was a fine gam we had…' H. Melville in Moby Dick); used to describe the social gatherings and interactions of courting albatrosses. Also a collective term for a group of whales.

GIS Geographic Information System A powerful computer mapping system for storage, analysis and dissemination of primary and derived information about the earth, in which all data is spatially referenced by geographic coordinates (e.g. north, east). In addition to primary data, such as climatic and soil characteristics, GIS can be used to calculate derived values, such as erosion hazard, forest yield class, or land suitability for specified land-use types. Data is usually derived from maps and values can be printed out as maps.

GPS Global Positioning System A satellite-based system of at least 24 orbiting transmitters maintained by the US Department of Defence for accurately locating three-dimensional coordinates anywhere on earth.

guano Dried bird droppings; large accumulated deposits, rich in nitrogen, have formed the basis of historic fertiliser 'mining' operations, especially on islands off Peru.

Hebe Large genus of evergreen flowering shrubs of New Zealand.

high seas All parts of a sea or ocean not included within the territorial boundaries of a state; designated as International Waters, freely open to all nations by UN Convention. In fisheries terms refers to waters outside a nation's *EEZ*.

interannual Between years.

IUCN The World Conservation Union (formerly International Union for the Conservation of Nature and Natural Resources) A multicultural, multilingual organisation based in Gland, Switzerland: uniting 83 states, 110 government agencies and 800 *NGOs*, its aims are to assist in conserving the integrity and diversity of nature and promote the ecologically sustainable use of natural resources.

longline fishery, longlining Fishing techniques employing a main line (up to 150 kilometres [93 miles] long) either anchored to the bottom or supported by floats horizontally at a predetermined depth, with baited hooks (up to 3000 per line) dangling on branch lines running off at regular intervals (0.4–50 metres [1.3–164 ft] long). Mainly they target *demersal* species (to 2500 metres [8202 ft] deep), although *pelagic* longlines are used to catch tuna closer to the surface. Longlines can be operated from virtually any size vessels and generally yield high-quality product. With many millions of hooks set annually (a single vessel may set and haul ±40,000 per day) the fishery has a reputation for being indiscriminate across a broad range of species and has been associated with high levels of seabird and other *bycatch*.

megaherb Any of an array of oversized herbaceous perennial plants with large leaves and abundant floral displays, adapted to the oceanic environments of New Zealand's subantarctic islands.

metabolism; metabolic rate The sum of the complex suite of basic biochemical reactions within an organism and its cells. These are processes that form the basis of life, the speed and efficiency of which determine how much food an organism requires to maintain its energy expenditure.

MPA Marine Protected Area (as per D. Hyrenbach) A marine area — intertidal or subtidal, within territorial waters, *EEZs* or in the *high seas* — set aside by law for the preservation and protection of important natural and cultural features, including marine biodiversity, fishery resources, specific habitats (e.g. mangroves, coral reefs) or species, and for the regulation of the scientific, educational and recreational uses of the seabed, the overlying water and the associated flora, fauna, historical and cultural features. Such areas may include nature reserves, wildlife sanctuaries, natural monuments, and cultural or archaeological reserves.

NGO non-governmental organisation Any national or international body, especially not-for-profit entities, involved in developmental or conservation activities, that are not associated with any local, state or federal government or United Nations organisation.

nuptial display In birds the ritualised physical movements and/or vocal calls, often exaggerated and repeated in elaborate sequences, important in courtship for forming a strong and lasting pair-bond prior to mating and nesting.

overfished Term describing a particular species or stock that has been exploited beyond a biological reference point, where its natural abundance (or biomass) is depleted to reproduction levels considered 'unsafe' to maintain the integrity of the species. Resultant species composition and dominance can have a domino effect throughout an *ecosystem*.

pelagic Living in or frequenting open oceans and seas, relating to those areas as opposed to waters close to land.

philopatry From the Greek 'home-loving', in animal behaviour it refers to the tendency of an individual to stay in or return to a specific location to feed or breed.

Procellariiformes *Taxonomic* order of the 'tube-nosed' birds — those characterised by having their nostrils enclosed in one or two tubes — containing four families representing all species of albatross, petrels, storm petrels, prions, shearwaters and diving petrels; all are highly *pelagic* and distributed throughout the world's seas to the highest latitudes. The order includes the smallest, the largest and most numerous seabirds. Derived from Latin procella, a storm, referring to the birds' ability to ride strong winds.

PTT Platform Transmitter Terminal A small satellite transmitter that can be attached to an animal or bird in order to monitor its movements or behaviour.

regime shift A rapid medium- or long-term change in environmental conditions, often climatic, from one relatively stable state to another that impacts on an *ecosystem* and its productivity.

RFMO Regional Fisheries Management Organisation An affiliation of nations formed to coordinate efforts to manage fisheries and to implement governing regulations, either for a particular species or for all living marine resources in a particular region.

RSPB Royal Society for the Protection of Birds A large UK-based charity *NGO* with over one million members, working to secure a healthy environment for birds and other wildlife.

salps Resembling jellyfish, they are planktonic, filter-feeding, barrel-shaped, gelatinous animals with complex life cycles, belonging to the tunicates of the phylum Chordata. Fast growing, they range in size from a few millimetres up to a metre (3 ft), can form large swarms and, though low in nutrition value, are a main food source for many fish and to a lesser extent seabirds and marine mammals.

SANAP South African National Antarctic Programme Administered through the Department of Environmental Affairs and Tourism, with bases in Antarctica, Marion and Gough Islands, it is South Africa's scientific research body whose mission is 'to increase understanding of the natural environment in the Antarctic and the Southern Ocean'.

SCAR Scientific Committee on Antarctic Research An interdisciplinary committee of the International Council for Science (ICSU). Charged with initiating, developing and coordinating high-quality international scientific research in the Antarctic region, it provides objective and independent scientific advice to the Antarctic Treaty Consultative Meetings and other organisations on issues of science and conservation affecting the management of Antarctica and the Southern Ocean.

scoria In geology/volcanology, the porous cinder-like fragments of cooled lava.

sessile In zoology, refers to organisms (e.g. barnacles) that are unable to move, i.e. those attached to the substrate on which they live.

Southern Ocean The contiguous deep, cold and tempestuous seas of the South Atlantic, South Pacific and southern Indian Oceans, surrounding Antarctica and generally spanning the Polar Front (*Antarctic Convergence*) to approximately 60 degrees south latitude, though commonly extended northwards to include all the waters south of New Zealand, Australia and South America and as far as 48 degrees south in the South Atlantic. Not bounded by any land mass, cyclonic storms travel continually eastwards, with the massive and highly productive Antarctic Circumpolar Current playing a crucial role in global ocean circulation and productivity.

spatially explicit In studies of population dynamics and ecological processes, it refers to a modern computational modelling technique useful for conservation biologists and resource managers in realistically predicting population and animal community responses to phenomena such as climate change, habitat modification, regional land and sea uses, or changes in local and regional management strategies. These are difficult to study with traditional ecological techniques. To project the impact of landscape change on wildlife populations, spatially explicit models incorporate habitat-specific information from field studies of life history, behaviour, *demographics*, distribution and dispersal of organisms, and thus relate explicitly to both the spatial and temporal complexities of the landscape in which the model subject exists.

stochastic, stochasticity In statistics and probability, a process involving chance or containing one or more random variables, or a process dependent on variable parameters; lacking any order or plan.

subantarctic Relating to the vast regions of the *Southern Ocean* north of Antarctica and south of temperate hydrographical zones. Dominated by strong westerly winds, the Subantarctic Zone is defined by the fluctuating boundaries where oceanic waters of different temperatures and salinity meet: between the Subantarctic Front at the northern reaches of the east-flowing Antarctic Circumpolar Current, and the Subtropical Front at 30–45 degrees south.

subtropical convergence Zone in the general vicinity of latitude 30–45 degrees south is the region of the Pacific, Atlantic and Indian Oceans where warmer water of tropical origin interfaces with colder waters of the *subantarctic* zone.

sulcus A distinctive feature of a bird's bill or beak, it is the fleshy lateral line or groove of the lower mandible occasionally referred to as a key to species identification.

sympatric In ecology, species, especially those closely related, occupying the same or overlapping geographic areas without interbreeding.

TAAF Terres Australes et Antarctiques Françaises The autonomous French overseas territories administration responsible for the islands of Saint Paul and Amsterdam, the Crozet and Kerguelen Archipelagos, and their *EEZs* and Antarctica's Adelie Land, as well as other Indian Ocean islands.

taxon (plural taxa), taxonomic A naturally related animal or plant group; the organised, ranked systematic category such as phylum, class, order, family, genus, or species in which an individual is classified according to its evolutionary relationships.

tori lines Relatively simple bird-scaring devices consisting of suspended colourful flapping streamers dangling behind a vessel, above the main fishing line or trawl, deployed to deter birds from approaching the dangerous area where baits or offal linger near the surface. Tori means 'bird' in Japanese; it has been shown that, depending on design, tori line use can reduce seabird mortality in pelagic fisheries by 30 to 70 per cent or more.

trawl fisheries Any of a number of techniques employing one or more vessels towing a funnel-shaped net through the water. Consisting of a cone-shaped body, extended at the opening by wings and ending in a bag (also known as the codend); different designs target different species at various depths, e.g. *demersal* or *pelagic* fish stocks.

vermicular Literally means 'having wormlike shape or motion'; in bird plumage refers to wavy markings of the feathers (vermiculation).

WWF World Wide Fund For Nature (formerly World Wildlife Fund) Since 1961 a global environmental conservation organisation now operating in more than 100 countries, it aims to stop the degradation of the planet's natural environment and conserve nature and ecological processes by preserving biodiversity, ensuring sustainable use of natural resources and promoting the reduction of pollution and wasteful use of resources and energy. (The name World Wildlife Fund is still in effect in North America.)

Further Reading

During almost six years compiling this book, our quest for the most intriguing and up-to-date information has taken us on routes seemingly as diverse and circuitous as the flight paths of the albatrosses themselves. Though published works are many, most are technical papers buried in the annals of scientific literature. Inevitably, in such a dynamic field as albatross research, new discoveries often render previous 'facts' obsolete, so even relatively recent material can appear out-dated, often leading to confusing or contradictory results. Even the species list has changed dramatically as DNA research led to a thorough taxonomic review in recent years, with 14 species now reorganised into 22. Throughout the book we have endeavoured to incorporate the latest information. However, while much of the text has been either reviewed or contributed by researchers and experts in their fields, any errors and omissions remain entirely our own.

If the pages of our book have piqued your interest, we would like to direct the inquisitive reader to begin his or her own venture into albatross realms. Space constraints prohibit inclusions of an exhaustive references and citations list, though some of the expert contributors in Part II include their own data sources. Following is a selection of titles we have referred to, chosen for their interest and breadth of information, that we feel offer a good starting point for individual research and further reading.

Brooke, Michael. *Albatrosses and Petrels across the World*. Illustrated by John Cox. New York: Oxford University Press, 2004.
del Hoyo, Josep, Andrew Elliott & Jordi Sargatal (eds). *Handbook of the Birds of the World. Volume 1: Ostrich to ducks*. Barcelona: Lynx Edicions, 1992.
Onley, Derek & Paul Scofield. *Albatrosses, Petrels and Shearwaters of the World*. London and New Jersey: Christopher Helm and Princeton University Press, 2007.
Peat, Neville. *Subantarctic New Zealand: A rare heritage*. Wellington: Department of Conservation Te Papa Atawhai, 2006.
Robertson, Graham & Rosemary Gales (eds). *Albatross: Biology and conservation*. Chipping Norton: Surrey Beatty & Sons, 1998.
Safina, Carl. *Eye of the Albatross: Visions of hope and survival*. New York: Henry Holt and Company, 2002.
Shirihai, Hadoram. *A Complete Guide to Antarctic Wildlife*. Illustrated by Brett Jarrett. London and New Jersey: A & C Black and Princeton University Press, 2008
Sibley, Charles G. and Burt Monroe. *Distribution and Taxonomy of Birds of the World*. New Haven: Yale University Press, 1990.
Terauds, Aleks & Fiona Stewart. *Albatross: Elusive mariners of the Southern Ocean*. Sydney: Reed New Holland, 2005.
Tickell, W.L.N. *Albatrosses*. Sussex and New Haven: Pica Press and Yale University Press, 2000.
'Tracking Ocean Wanderers: The global distribution of albatrosses and petrels'. *Results from the Global Procellariiform Tracking Workshop*, 1–5 September, 2003, Gordon's Bay, South Africa. Cambridge: Birdlife International, 2004.

Websites with specific information on albatrosses:
www.acap.aq
www.birdlife.org/action/campaigns/save_the_albatross/index.html
www.forestandbird.org.nz/Marine/albatross
www.rspb.org.uk/supporting/campaigns/albatross/index.asp
www.savethealbatross.net

Index

Adams Island (Auckland Islands) 4–5, 32, 33, 36, 38, 80, 83, 190, 191, 194, 195, 201, 207
aerial displays 98, 99, 103, 106, 118, 189, 191, 230, 231, 232 see also flight, light-mantled albatross, sooty albatross
Africa 218, 219
Age of Discovery 139
Agreement on the Conservation of Albatrosses and Petrels (ACAP) 149, 177, 179
aggression and defence displays 189
Agulhas Current 157
ahodori ('fool bird') 186, 225
Alaska 147, 162, 163, 226, 233
Albatross Island (South Georgia Islands) 25, 27, 28, 79
Albatross Island (Tasmania) 16–17, 79, 150, 154, 190, 208, 209
Albatross Task Force 174, 175, 177
Aleutian Islands 143, 162, 163, 222, 226, 233
Alfaro, Joanna 161
Amey, Jacinda 87
Amlia Island (Aleutian Islands) 162
Amsterdam albatross 16–17, 27, 37, 148, 149, 151, 158, 159, 186, 190, 197–198 (species profile), 233; population increase 158–159
Amsterdam Island 16–17, 37, 70, 141, 149, 158, 159, 190, 191, 197, 198, 219, 220, 221, 231, 232, 233
Anisotome megaherbs 30; A. latifolia 100 see also megaherbs
ancestral peoples 143 see also iwi, Maori, Moriori
Annekov Island (South Georgia Islands) 203
Antarctic: Convergence 16–17, 97, 229; expeditions 139; Peninsula 58
Antarctica 100, 183, 195, 233
Antipodean albatross 4–5, 16–17, 21, 25, 27, 28, 29, 30, 31, 32, 33, 36, 38–39, 144, 148, 150, 151, 155, 182, 186, 187, 190, 193–195 (species profile); chick 31, 32; courtship 30, 32, 38–39
Antipodes Island 16–17, 28, 29, 30, 35, 190, 191, 193, 194, 195, 203, 204, 207, 229, 230
Apotres Island (Crozet Islands) 191
Argentina 156, 165, 175, 176, 177, 200, 218
Argos weather satellite system 164, 165, 169 see also satellite tracking
artisanal fisheries 146, 149, 161, 169, 176, 228 see also fishing and fisheries
Atlantic Ocean 59, 70, 119, 156, 158, 159, 203, 204 see also North Atlantic, South Atlantic
Atlantic yellow-nosed albatross 2–3, 16–17, 26, 62, 63, 70, 72, 73, 74, 75, 76, 149, 150, 151, 171, 177, 187, 191, 217–219 (species profile), 233; colonies 74; gamming groups 74, nests 72, 73, 74, 75
Auckland Islands 16–17, 32, 33, 46, 79, 80, 81, 99, 150, 154, 190, 191, 193, 195, 201, 207, 229, 233 see also Adams Island, Disappointment Island, Enderby Island
Australia 29, 70, 79, 97, 140, 143, 150, 151, 156, 157, 158, 159, 164, 179, 180, 188, 195, 196, 205, 208, 209, 216, 218, 220, 233
avian cholera 158, 159, 198, 219, 221, 232
avian pox virus 209, 224

bands, banding 85, 121, 154, 155, 157, 158, 161, 164, 166, 178, 202, 228
Banks, Joseph 143
Bay of Isles, South Georgia 156
Beagle HMS 21, 143

Beagle Channel 58
Beauchêne Island (Falkland Islands) 53, 54, 56, 203, 204
Beck, Rollo 199
Beebe, William 147
Benguela Current 157
Bering Sea 162, 163, 226
Bermuda 119, 143
Bird Island (South Georgia) 143, 156, 157, 190, 191, 192, 203, 213, 214
BirdLife International 14, 56, 148, 156, 157, 174, 175, 178, 179, 186, 207; Global Procellariform Tracking Database 175; Global Seabird Programme 156, 174
bird-scaring lines (streamer lines) 175, 177 see also tori lines
bird-strike 113, 145
black-browed albatross 16–17, 52–59, 70, 72, 140, 141, 148, 149, 151, 152, 153, 155, 156, 157, 158, 159, 175, 176, 177, 180, 181, 183, 187, 189, 190, 202–204 (species profile), 233; chick 54; colonies 56, 157; courtship 55, 60–61; decline of 55–56; population trends 159; range 58–59
black-footed albatross 16–17, 115–119, 120, 121, 145, 149, 150, 151, 162, 163, 180, 181, 186, 191, 223–224 (species profile), 233; chick 116, 117; courtship 118, 122–123
Blechnum palmiforme 62, 219
Bligh, Captain William 83
blowflies 200
Bollons Island (Antipodes Islands) 191, 204, 207
Bonin Islands (also known as Ogasawara Islands) 16–17, 163, 191, 221, 222, 223, 226
Bonn Convention on Migratory Species 179
boobies 20, 228
Bounty HMS 83
Bounty Islands 16–17, 78, 83, 85, 86, 87, 88, 89, 144, 155, 191, 209, 210, 211
Bouvet Island 16–17
Brazil 156, 157, 175, 176, 177, 196, 200, 218; Current 177; waters 177
breeding 153, 156, 188–189, 190; colonies, new 155; locations 16–17, 189; pairs 151; populations, estimating 187, 188
Brewster-Sanford, Rollin 199
British Antarctic Survey 157
brooding 188
Brothers, Nigel 156, 158
Buller, Sir Walter Lawry 215
Buller's albatross 16–17, 46, 47, 64, 65, 68, 136–137, 147, 149, 150, 151, 154, 166–167, 172, 176, 187, 188, 189, 190, 215–217 (species profile), 233; banding chicks 166; breeding 166; chick 66, 166; colony 64; courtship 76–77; nests 65, 66; population dynamics 166; sex ratio of chicks 167; southern Buller's subspecies 166; tracking 166
bunting, Gough Island 104
bycatch 145, 147, 148, 156, 157, 159, 163, 164, 168, 169, 173, 175, 176–177, 193, 211, 217; mitigation 56–57, 172, 173, 174, 175, 177, 179, 181, 182, 183

California 116, 152, 181, 224; Current 180
Campbell albatross 16–17, 53, 59, 60, 101, 149, 150, 151, 187, 188, 204–206 (species profile); chicks 58, 59
Campbell Islands 16–17, 20, 41, 42, 43, 44, 45, 46, 50, 51, 60, 68, 69, 97, 98, 99, 101, 108, 147, 155, 186, 190, 191, 193, 194, 201, 202, 203, 204, 205, 213, 214, 229, 230, 233

Campbell Plateau 206
Canada 163, 226
Cape Alexandra, South Georgia Island 157
Cape of Good Hope 165
Cape Horn 21, 69, 141, 164
Cape sheep 192
Cape Town 175, 179, 183
caracara, striated 54, 55, 204
carrion 193, 195, 230, 232
Carter, Tom 219
cattle 198, 232
cats, feral 198, 223, 230, 232
casuarina trees 113
cephalopods 193, 195, 198, 200, 202, 206, 207, 209, 211, 213, 217, 219, 222, 226, 230, 232 see also squid
Chatham albatross 16–17, 22–23, 24, 79, 90–93, 132, 149, 168–169, 151, 155, 176, 187, 206, 208, 209, 211–213 (species profile); adult mortality 168; banding studies 168, 169; breeding population 168; chick 91; courtship 90, 94–95; fledging 168; GPS loggers 169; heat stress 169; hatching 168; migration 169; nest 90, 91, 92, 93; non-breeders 90
Chatham Island button daisy (Leptinella featherstonii) 48, 200
Chatham Islands 16–17, 47, 49, 65, 67, 70, 90, 139, 143, 149, 164, 165, 166, 168, 169, 191, 199, 200, 210, 211, 212, 213, 215, 217, 218
Chatham Rise 168, 212
chicks 153, 154, 159, 188; chick mortality 158; projectile (defensive) vomiting 66, 82, 166 see also individual species
Chile 58, 69, 155, 168, 169, 173, 175, 176, 177, 195, 200, 203, 204, 214; longline fishing fleet 176
Chilean seabass see Patagonian toothfish see also fish
China 145
circumnavigation 157, 165
Clarion Island (Revillagigado Islands) 191, 222
climate change 91, 169
Coleridge, Samuel Taylor 20, 142, 143, 186
colonies: decline of 55, 56, 102, 146, 148–151 see also individual species
Commission for the Conservation of Antarctic Marine Living Resources (CCAMLR) 175
Continental Shelf 180, 181, 183, 191, 222, 226 see also Patagonian Shelf, South American Shelf
Cook, James 141, 143, 204
Cooper Island (South Georgia Islands) 203
Cordell Bank National Marine Sanctuary 181
cormorants 130, 178 see also shag
courtship 189 see also individual species
Crozet Islands 16–17, 29, 70, 141, 149, 152, 153, 154, 155, 158, 159, 183, 190, 191, 192, 210, 214, 220, 231, 233
crustaceans 193, 196, 198, 200, 202, 209, 213, 217, 219, 220, 222, 224, 226, 230, 232
Cumberland Bay, South Georgia 99
da Cunha, Tristão 139, 170
curlew, bristle-thighed 223, 224
cuttlefish, giant 195

Darwin, Charles 21, 143
deep-sea longlining 183; trawlers 56, 173 see also fishing and fisheries
demersal (bottom) longline fisheries 175, 177 see also fishing and fisheries
Department of Conservation (DOC) (New Zealand) 32, 83, 168
detection loggers see satellite tracking

Delvolvé, Laure 145
Dias, Bartolomeu 139
Diego de Almagro Island 16–17, 190, 203
Diego Ramirez Islands 16–17, 69, 141, 189, 190, 191, 203, 214
Diomedea (genus) 20, 186, 188, 189, 190
Diomedea amsterdamensis see Amsterdam albatross
Diomedea antipodensis see Antipodean albatross
Diomedea dabbenena see Tristan albatross
Diomedea epomophora see southern royal albatross
Diomedea exulans see wandering albatross
Diomedea gibsoni see Gibson's albatross
Diomedea sanfordi see northern royal albatross
Diomedeidae (family) 144, 148, 186
Diomedes 20, 186
Disappointment Island (Auckland Islands) 38, 79, 80, 90, 191, 194, 206, 207
disease 149, 159 see also avian cholera, avian pox virus
dispersers 154, 155
distribution 16–17, 180, 181
DNA studies 30, 186 see also genetics
dogs 223
dolphins 228; spinner 114
Dracophyllum 202
Drake Passage 141, 183
driftnet fishing 146, 149, 222, 224 see also fishing and fisheries
Duke of Portland 143
Dumas, Alexander 144
Dundonald 79
Dunedin, New Zealand 47, 155, 164
dynamic soaring 25–27, 68, 144, 152–153 see also flight

Earle, Augustus 170
East China Sea 143, 225
Eastern Island (Midway Atoll) 147, 154, 222, 223
Ecuador 125, 150, 176, 227
Edinburgh of the Seven Seas, Tristan da Cunha 34, 171
egg trade 112
El Niño 130, 132, 168
Enderby Island (Auckland Islands) 46, 190, 201, 202
environmental change 158, 159
Erysipelas bacteria 232
Erysipelothrix rhusiopathiae 198, 219, 221
Española Island (Galapagos Islands) 124, 125, 126, 128, 130, 131, 132, 133, 135, 160, 161, 180, 191, 227, 228, 233
etymology 186
Euphausia superba 204 see also krill
Exclusive Economic Zones (EEZs) 180, 181
Eye of the Albatross 20

Falkland Islands 16–17, 53, 55, 56, 57, 156, 176, 177, 190, 202, 203, 204, 233; fisheries 56
Falklands Conservation, BirdLife International 56
Faroe Islands 155
feather trade 111, 112, 119, 120, 145, 148, 163, 208, 224, 226
Feynman, Richard 146
fernbird 63
fighting 87, 160, 189
fire 221
Fischer, Karen 162
fish 56, 166, 193, 195, 196, 198, 200, 202, 204, 207, 209, 211, 213, 215, 217, 219,

220, 222, 226, 228, 230, 232; hake 176, 177; kingclip 177; mackerel 207, 209; Patagonian toothfish 176, 177, 183; redbait 207; salmon 226; southern blue whiting 206; swordfish 175, 176; tuna 56, 156, 175; yellownose skate 177 *see also* krill, squid
fish eggs 116, 117, 222, 224
fishing and fisheries 37, 56, 58, 79, 102, 132, 145–146, 148, 150, 156, 157, 158, 163, 169, 172–173, 174–175, 180, 196, 198, 200, 202, 206, 211, 213, 215, 217, 219, 220, 222, 226, 228: Argentinian 177; Indian Ocean southern blue-fin tuna 158; Japanese tuna 156; jiggers 177; methods 172; mortality from 179; night-setting 172, 175, 177; operations 150; monitoring 181; observers 175; pirate 146, 182; practices 176; seabird-smart techniques 173; South American 176–177; tuna 157, 209 *see also* artisanal fisheries, bird-scaring lines, bycatch, driftnet fishing, longline fisheries, tori lines, trawl fisheries
flight 152–153; aspect ratios 152; dynamic soaring 25–27, 68, 144, 152–153; gliding 153; landing 60, 61, 153; take-off 153 *see also* aerial displays, dynamic soaring, wings *and individual species*
Food and Agriculture Organisation (FAO) 145, 178
Forty-Fours, The (Chatham Islands) 46, 47, 48, 49, 65, 139, 147, 165, 166, 190, 199, 200, 216
fossils 119, 143
France 70, 151, 158
French Frigate Shoals (North-western Hawaiian Chain) 222, 223
French Southern Territories 16–17, 158
fulmar: northern (also known as Arctic) 162, 186
fur seals 28, 49, 79, 83, 86, 87, 143, 144, 156, 200; disturbance to albatrosses 156, 193, 211

Galapagos Islands 16–17, 125, 126, 130, 147, 149, 160, 176, 228, 233
Galapagos Marine Reserve 132, 161, 180, 181, 228
Galapagos waved albatross *see* waved albatross
gamming 30, 42, 74, 139, 185, 189
gannets, Australasian 209; northern 59, 153, 155
genetics 154, 155, 161, 166, 186 *see also* DNA studies
Genovesa Island (Galapagos Islands) 227
geo-location loggers *see* satellite tracking
giant tortoise, Galapagos 161
Gibson, J.D. 'Doug' 156, 186, 193
Gibson Plumage Index 187, 193
Gibson's albatross, subspecies of Antipodean albatross 186, 187; courtship 38–39 *see also* Antipodean albatross
Glass, James 33, 171
Glass, Simon 171
global warming 56, 213 *see also* climate change
goats, feral 170, 228
golden gooney *see* short-tailed albatross
goney 186
Goneydale 34
gonia 171
gony or goony 186
gooney birds 144, 186
Gough Island 16–17, 33, 34, 35, 70, 73, 74, 75, 76, 96, 100, 102, 103–106, 107, 147, 149, 176, 178, 179, 190, 191, 196, 197, 217, 218, 219, 230, 231
GPS tracking logger 69, 165 *see also* satellite tracking
Grand Jason Island (Falkland Islands) 203
great albatrosses 186–187, 189
Great Australian Bight 202
grey-headed albatross 16–17, 68–70, 141, 149, 150, 151, 155, 156, 157, 158, 176, 180, 181, 187, 188, 189, 191, 213–215 (species profile), 233; chick 69; circumnavigation 69; colonies 69; nest 69; satellite tracks 157
Guadalupe Island 16–17, 191, 221, 222
Guano Islands (Peru) 130
guano mining 112, 148
Gulf of Alaska 163
gull, red-billed 217

hailstones 43, 202
hake *see* fish
Hasegawa, H. 226
Hawaii 120, 145, 152, 155, 163, 181, 223, 224, 226, 233
hawk, Galapagos 133, 228
Heard Island 16–17, 147, 154, 155, 190, 191, 229
heart rates 153
heat and heat stress 161, 169
Henry B Paul 171
Hermaness, Scotland 155
Hermes Reef (North-western Hawaiian Chain) 222, 223
Hokkaido 224
homing ability 154
Hough, Ada 47
Howard, Charles 140
human: early encounters with 139–143, 170; predation 74, 79, 132, 143, 144, 146, 148, 161, 163, 170–171, 200, 204, 213, 217, 219
Humboldt Current 130, 132, 135, 160, 176
Hurley, Frank 139
Hutton, Captain 144
Hydrobatidae 144

iguana, marine 160
Ildefonso Islands 16–17, 190, 203
Île aux Cochons (Crozet Islands) 192
Île des Pingouins (Crozet Islands) 155, 191
immatures 189 *see also individual species profiles*
Inaccessible Island (Tristan da Cunha Islands) 147, 190, 191, 197, 218, 219
incubation 166, 188 *see also individual species*
Indian Ocean 37, 59, 70, 97, 139, 140, 141, 155, 156, 157, 158, 159, 164, 178, 182, 183, 195, 196, 198, 207, 209, 210, 214, 218, 220, 231, 232, 233
Indian yellow-nosed albatross 16–17, 70, 71, 149, 150, 151, 155, 158, 159, 187, 191, 219–221 (species profile), 233; chicks 158; colonies 71; population trends 159
International Ornithological Congress, 1998 165
International Plan of Action for Reducing Incidental Catch of Seabirds in Longline Fisheries 179
introduced/alien species: plants 113; predators 179 *see also* cats, dogs, goats, mice, pigs, rats
iwi 47 *see also* Maori, Moriori

Japan 111, 119, 120, 143, 145, 147, 151, 155, 163, 223, 224, 233
Jeanette Marie Island (Campbell Island) 190, 205
jiggers 177 *see also* fishing and fisheries
Jouventin, P 186
juveniles *see* immatures

Kaikoura Peninsula (New Zealand) 21, 30, 31, 144, 233
Kamchatka Peninsula (Russia) 226
Kauai Island (Hawaii) 222
Kaula Island (North-western Hawaiian Chain) 223
Kerguelen Islands 16–17, 70, 141, 150, 154, 155, 158, 166, 183, 190, 191, 192, 214, 220, 229, 232, 233
Kipling, Rudyard 144
krill 159, 204, 215
Kure Atoll (North-western Hawaiian Chain) 222

Kuril Islands 226

La Plata Island (also known as Isla de la Plata) 16–17, 125, 191, 227
Laysan albatross 16–17, 110–116, 147, 150, 151, 154, 155, 162, 163, 186, 221–223 (species profile), 233; chicks 114, 115; colonies 112; courtship 122–123; nests 114; non-nesters 115
Laysan Island (North-western Hawaiian Chain) 145, 191, 222, 223
lead poisoning 145, 223
Leeward Islands (Hawaii) 145
light-mantled albatross 16–17, 72, 97–100, 108, 150, 151, 158, 159, 189, 191, 229–230 (species profile); chick 99, 101; courtship 99, 109; incubation 100; nest 100, 101
Lion Island (Bounty Islands) 85
Lisianki Island (North-western Hawaiian Chain) 222, 223
Little Sister Island (Chatham Islands) 164, 165
longevity 188 *see also individual species profiles*
longline fishing and fisheries 56, 146, 148, 149, 150, 156, 157, 158, 159, 168, 171, 172, 174, 175, 176, 177, 178, 179, 181, 182, 183, 195, 202, 204, 207, 209, 219, 222, 224, 230, 232; artisanal 169; demersal 211; inshore 173; pelagic (open sea) 175; Peru 176; Uruguay 177

MacDonald Island 16–17, 147, 158, 183, 190
mackerel *see* fish
Macquarie Island 16–17, 100, 147, 154, 190, 191, 214, 229, 230, 232
Macquarie Island Marine Park 180, 181
Magellan, Ferdinand 140
Mahalia, sailing yacht 46, 47, 83, 84, 87, 97, 147
Malvinas *see* Falkland Islands
Malvinas Current 177
Mangel, Jeffrey 161
Maori 143, 168
Marine Protected Areas 145, 177, 180, 181
marine reserves 181 *see also* Galapagos Marine Reserve
marine zoning 181
Marion Island (Prince Edward Islands) 141, 178, 191, 192, 214, 229, 230, 231
megaherbs 30, 41, 51, 79, 100 *see also Anisotome*, *Pleurophyllum*
Mewstone 16–17, 79, 147, 190, 208
Mexico 222
mice, predation on albatross chicks 35, 147, 149, 179, 197, 219
Mid-Atlantic Ridge 171
Middle Sister Island (Chatham Islands) 165
Midway Atoll (North-western Hawaiian Chain) 111, 112, 113, 118, 119, 120, 121, 122, 145, 147, 154, 155, 181, 191, 222, 223, 224, 226, 233
migration 149, 156, 157, 165
mockingbird, Española 228
mollymawks 186, 187, 188, 189, 190 *see also Thalasarche, individual species profiles*
Monument Harbour, Campbell Island 97, 98, 100, 101
Moriori 143, 168
mortality 156, 157, 174, 175, 176–177, 179
mosquitoes 130, 161
Mount Honey (Campbell Island) 45, 98
Mount Lyall (Campbell Island) 186, 202
Mukojima (Bonin Islands) 222
Murphy, Robert Cushman 178

Namibia 175, 207, 209
Nansen, Fridtjof 147
natal island 30 *see also* philopatry
National Marine Sanctuaries 180
Natural Resources Department (Tristan da Cunha) 34, 171
navigation 154
Necker Island (North-western Hawaiian Chain) 223
Neruda, Pablo 177

nests 54, 58, 166, 188–189; pedestal 59, 65, 66
New Island (Falkland Islands) 55, 57, 203
New South Wales 156, 198
New Zealand 20, 21, 30, 32, 35, 41, 47, 59, 63, 70, 79, 97, 132, 140, 143, 144, 149, 150, 151, 154, 155, 164, 165, 166, 168, 172, 173, 176, 177, 182, 188, 194, 195, 199, 200, 201, 202, 204, 205, 206, 207, 209, 211, 215, 216, 218, 220, 233; Department of Conservation (DOC) 32, 87, 168; fisheries 56; Ministry of Fisheries 168; subantarctic islands 83, 147, 166, 232 *see also* Antipodes Island, Auckland Islands, Bounty Islands, Campbell Islands, The Snares
New Zealand Geographic Society 144
Nicholls, David 164, 165
Nightingale Island (Tristan da Cunha Islands) 63, 72, 74, 77, 191, 218, 219
Nihoa Island (North-western Hawaiian Chain) 223
Niihau Island (Hawaii) 222
nomenclature 186 *see also* taxonomy
non-breeders 189
North America 143, 224
North Atlantic 143, 155
North Cape, Campell Island 59, 60, 68
North East Island (The Snares) 64
North Pacific 111, 114, 117, 119, 143, 145, 149, 150, 163, 181, 222, 223, 224, 225
North West Cape, Auckland Island 83
northern albatrosses (*Phoebastria*) 186, 189, 191
northern Buller's albatross 65–68, 168, 216 *see also* Buller's albatross
northern royal albatross 14, 16–17, 46–49, 66, 148, 149, 151, 152, 154, 164, 165, 168, 176, 186, 188, 189, 190, 199–200 (species profile), 233; chick 48; satellite tracking 48, 164
North-western Hawaiian Chain 16–17, 111, 221, 222, 223

Oahu Island (Hawaii) 221, 222
octopus 200
offal 196, 198, 207, 209, 228
Ogasawara Islands *see* Bonin Islands
Olearia: forest 64; *O. lyalli* 63, 65, 167
Otago Peninsula, New Zealand 47

Pacific albatross *see* Buller's albatross
Pacific Ocean 97, 132, 140, 150, 157, 158, 210, 212 *see also* North Pacific, South Pacific
Pacific Rim 226
pair bond 32, 189 *see also individual species*
parakeet 30
Patagonia 16–17, 56, 143, 164
Patagonian Shelf 16–17, 48, 146, 165, 176, 177, 190, 199, 202 *see also* South American Shelf
Patagonian toothfish 176, 177, 183 *see also* fish
Pearl Island (North-western Hawaiian Chain) 222, 223
Pedra Branca Island 16–17, 79, 147, 190, 208, 209
pelagic fishing 176; longline fisheries 175, 204 *see also* fishing and fisheries
Pelecanoididae 144
pelicans 130
penguins: 55, 84, 86, 141, 230: as food 230, 232; African (black-footed) 178; emperor 161, 174, 228; erect-crested 83, 84, 86, 87; king 161, 228; rockhopper 53, 55, 56, 70, 98; Snares crested 63
Perseverance Harbour, Campbell Island 45, 46
Peru 130, 132, 135, 145, 146, 149, 160, 161, 168, 169, 173, 176, 200, 210, 212, 214, 216, 227, 228; longliners 176
petrels 68, 72, 80, 144, 146, 157, 177, 179, 230; Atlantic 219; diving 63, 104, 144; giant 171, 175, 177, 179, 186, 207, 209, 211, 221, 230, (as food) 232; mottled 63;